PLAIN PLANE GEOMETRY

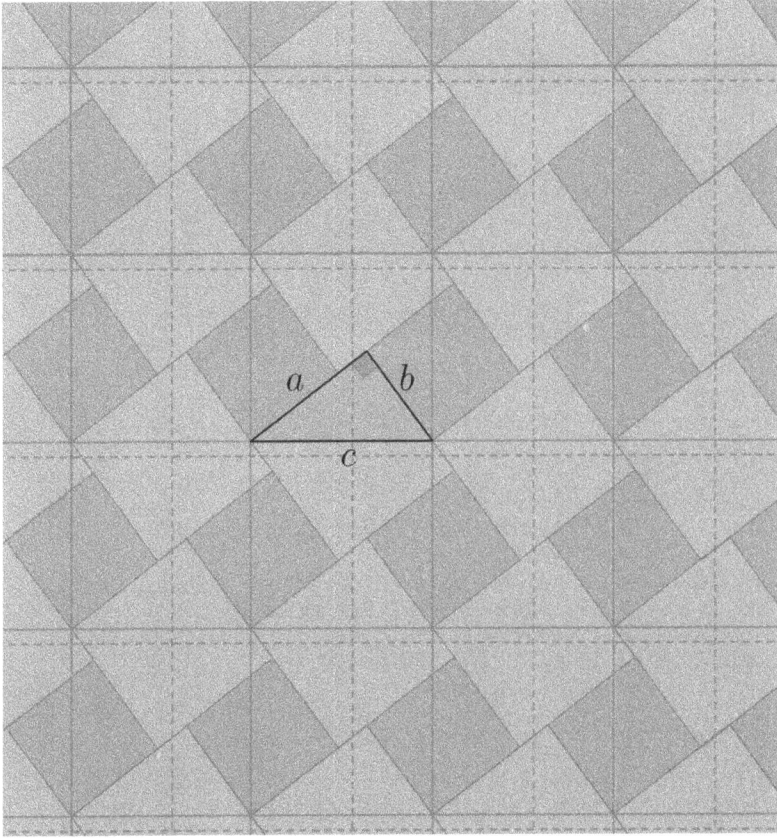

PLAIN PLANE GEOMETRY

Amol Sasane

London School of Economics, UK

World Scientific

NEW JERSEY · LONDON · SINGAPORE · BEIJING · SHANGHAI · HONG KONG · TAIPEI · CHENNAI · TOKYO

Published by

World Scientific Publishing Co. Pte. Ltd.

5 Toh Tuck Link, Singapore 596224

USA office: 27 Warren Street, Suite 401-402, Hackensack, NJ 07601

UK office: 57 Shelton Street, Covent Garden, London WC2H 9HE

Library of Congress Cataloging-in-Publication Data

Names: Sasane, A. (Amol), 1976–

Title: Plain plane geometry / Amol Sasane.

Other titles: Plane geometry

Description: New Jersey : World Scientific, 2016. | Includes bibliographical
   references and index.

Identifiers: LCCN 2015039584 | ISBN 9789814740432 (hardcover : alk. paper) |
   ISBN 9789814740449 (softcover : alk. paper)

Subjects: LCSH: Geometry, Plane--Study and teaching (Secondary) | Geometry,
   Plane--Study and teaching (Higher) | Geometry--Study and teaching
   (Secondary) | Geometry--Study and teaching (Higher)

Classification: LCC QA455 .S24 2016 | DDC 516.22--dc23

LC record available at http://lccn.loc.gov/2015039584

British Library Cataloguing-in-Publication Data

A catalogue record for this book is available from the British Library.

Cover design by Malin Christersson and Amol Sasane.
A proof of Pythagoras's Theorem by tessellation.

To Arun

Preface

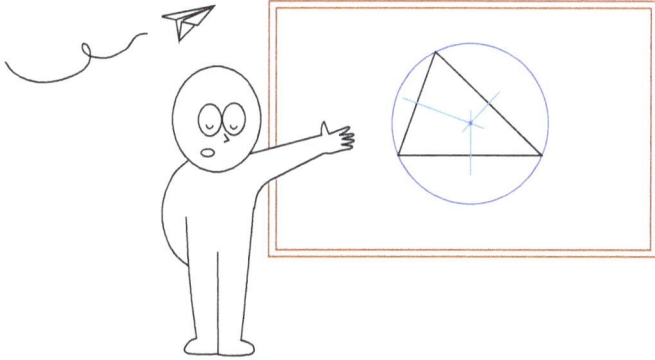

Why this book?

Because I want to share the feeling that planar geometry can be fun. Although as high school students, my generation were taught Euclidean plane geometry, it is common to encounter a modern-day high school mathematics textbook which is largely devoid of any pretty geometry theorems with proofs, as other important topics have replaced such old-fashioned things. The present book hopes to remedy this in some measure.

What's so special about geometry? We'll elaborate this below, but essentially:

(1) It showed that mathematics involves not just *numbers*, but *pictures* (which is how many mathematicians *think* when creating new maths).
(2) It convinced the student that mathematics is beautiful and not boring. On the contrary, doing Mathematics can be *enjoyable*!
(3) Besides all these lovely things, it taught students what a "proof" is, and prepared high school graduates for university level mathematics.

The aim of this book is to cover the basics of the wonderful subject of Planar Geometry, at high school level, requiring no prerequisites beyond arithmetic, and hopefully to convey the sense of joy which I had when I was taught geometry.

While one might argue about the use of teaching such outdated things, and contrast this with teaching other useful things such as Cartesian geometry, algorithms and so on, surely, as mentioned in item (3) above, there is one feature of Euclidean plane geometry which makes learning it worthwhile: it trains the student in understanding proofs, and also in devising one's own proofs: in other words, it inculcates the very spirit of Mathematics!

What is Planar Geometry?

Planar figures are figures in the plane, such as triangles, quadrilaterals and circles.

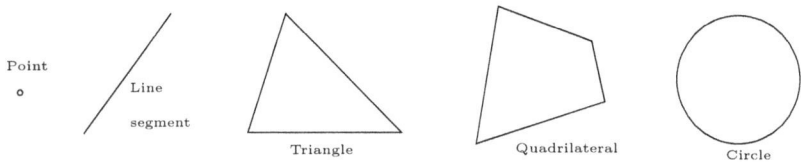

By *Planar Geometry*, we mean a study of planar figures and their geometrical properties. By a *geometric property* of a planar figure, we mean one which doesn't change under a "rigid" motion (that is motion that does not distort the figure: examples of such motion are rotation, translation and reflection). Examples of geometric properties that don't change are distances and angles.

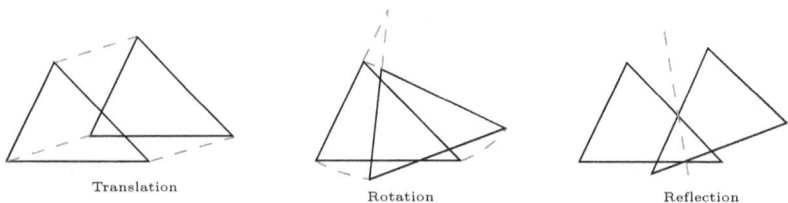

Why study Planar Geometry?

Planar Geometry in some sense marks the birth of Mathematics as it is practiced in modern times. Often it is considered as the first historical treatise (the works of Euclid) in which the rules of doing Mathematics were set out in a systematic manner. If one asks: What is Mathematics? Then the following is a rough answer:

> Mathematics is a subject in which we make up definitions and prove things about the defined objects in a logical manner.

Thus Mathematics can be likened to a game, like Chess, in which there are:

(1) objects (the chess board, chess pieces),
(2) rules of the game (how the chess pieces move etc.),
(3) the play (when two players actually play the game).

In a similar manner in Mathematics, there are:

(1) objects (definitions of mathematical objects),
(2) rules (mathematical logic),
(3) the play (making up theorems, that is, true statements about the objects and proving these truths).

This endeavor of doing mathematics is brought out very clearly for a school student by studying Planar Geometry à la Euclid, since

(1) The objects in Planar Geometry are points, lines, angles, triangles, quadrilaterals and circles, and these are quite easy to understand since they are concrete, and visual.
(2) The rules of the game are self-evident truths which are easy to accept, since they are again visual. These rules are also just a few in number and can be stated succinctly. Thus the logical structure of the proof becomes extremely clear to the student, when one gives reasons for the steps in the proof (such as usage of the Parallel Postulate, SAS Congruency Rule etc., which we will soon learn).
(3) The theorems in Planar Geometry, and their proofs are particularly beautiful and are ideal to convey the beauty of Mathematics to the uninitiated person. Indeed, prior to studying geometry, a school student's exposure is mostly limited to arithmetic, mostly with no "soul" or sense of "play". This changes drastically with Planar Geometry, since everything is visual, so the mind's eye can "see", and also play (by drawing and trying this and that, guessing, experiencing an "aha!"

moment, etc.). For example, in Exercise 3.27, we will encounter the following "proof without words", of Pythagoras's Theorem, saying that the square of the length of the biggest side in a right angled triangle is equal to the sum of the squares of the other two sides.

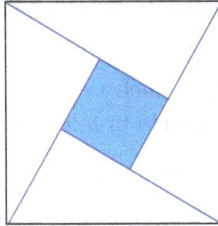

What will we learn in this book?

There are 5 chapters in the book:

(1) Geometrical figures
(2) Congruent triangles
(3) Quadrilaterals
(4) Similar triangles
(5) Circles

Perhaps the reader has encountered some or all of these objects before, and so might have some feeling of what is in store while reading this book. But besides this seemingly innocent backdrop of learning about planar geometry, the book has some ulterior motives:

(1) to learn drawing pictures, and realizing that Mathematics is not just about numbers, but can be very visual with pictures;
(2) that Mathematics can be fun: rather than being a monotonous process of carrying out some algorithm (like long division, calculation of percentages or solving a quadratic equation), it can be a creative process (where in order to construct a mathematical proof, one has to try out different possible ideas, get inspiration out of the blue, and sense exhilaration at solving an interesting problem (=puzzle) oneself, akin to solving an entire crossword puzzle, or winning a chess game, etc.;
(3) to teach what constitutes a proof (and en route instill rules of mathematical logic such as proof by contradiction, equivalent statements, converse of a statement, necessary and sufficient conditions, etc.);

(4) to develop mathematical maturity, which can be a combination of all of the above and more: when one knows what it means to "do Mathematics";

(5) to learn problem solving by doing the exercises, in which one can develop solution strategies. For example, learning to understand a mathematical problem by writing what is given and what is asked, finding a possible solution technique by asking if there is a simpler related problem which can be solved, or experimenting with the given data in the problem, considering extreme cases, and so on. Making informed and useful *guesses* in the form of possible constructions, or for example making claims about the equality of some pertinent angles or lengths or about perpendicularity in the picture, etc. Learning to subconsciously recognizing when one gets the key idea solving the problem, and then planning and writing the solution in a logical manner.

Funnily enough, no effort will be needed to specifically devote any energy on the above aims, as these will be *automatically* imparted when one practices Geometry!

How did Planar Geometry arise?

The word "geometry" is derived from the Greek word "geo" meaning "Earth", and "metron" meaning "measurement".

$$\text{Geo} + \text{metron} = \text{Geometry}.$$

Historically Geometry seems to have arisen from a need to measure land. This is imaginable, since if a flood swept away the demarcations between plots of cultivable land, there would have been a need to redraw boundaries, and presumably this could have given rise to specific problems in planar geometry, leading to advancement in the subject over time. Other reasons could have been architecture and astronomy.

Historians of Mathematics claim that while there's evidence to show that ancient civilizations, such as the Indus Valley Civilization, the ancient Egyptians, and the Babylonians, knew geometry, the subject of Planar Geometry was developed *systematically* by the Greeks, and culminated around 300 BC with the works of Euclid, called *Elements*, in thirteen volumes. In these works, a new manner of thinking was introduced in geometry, in which geometrical results were proved by starting from an initial small collection of self-evident assumptions called *axioms*: for example when a line intersects two parallel lines, then the corresponding angles are equal:

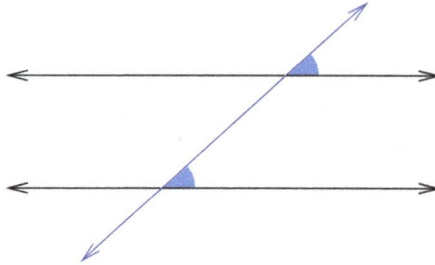

Then other results in geometry, such as Pythagoras's Theorem, were deduced from these fundamental axioms by a process of logical reasoning. These derived new truths were called *theorems*, and the body of reasoning associated with a theorem was called its *proof.*

Who is the book for?

The book is written at high school level or beginning university level, but can be read by anyone interested in the subject.

How should the book be read?

The book follows the "Definition-Theorem-Proof-Exercise" format of Mathematics. The exercises are an integral part of studying this book. They are a combination of elementary ones (meant for understanding the definitions or simple usages of the theorems learnt), and those which are more challenging (sometimes indicated by an asterisk (*)), involving a careful and elaborate use of the theorems. The student should feel free to skip exercises which seem particularly challenging at the first instance, and return back to them now and again. Occasionally, the exercises also treat topics which are discussed subsequently in the text, and so they can be postponed until the relevant stuff has been covered. Although detailed solutions are provided, the student should not be tempted to consult the given solution too soon. To this end, a section on helpful hints is included, in order to give a nudge in the right direction. However, the solutions stem from the author's own point of view, and the reader might have an alternative solution.

The book does not aim to be encyclopedic, and there are several omissions. Nevertheless the hope is that, with the "Less is more ($\cdots$ provided it is pretty $\cdots$)" dictum I have tried to follow in this book, the uninitiated reader enjoys this book to a sufficient degree in order to get hooked, and learn more!

Acknowledgements

I have made use of material from many diverse sources such as mathematics olympiad problem sets, mathematics education journals such as *The Mathematical Magazine* and *The Mathematical Gazette*, and online resources on topics in geometry. These are listed in the bibliography and at the end of chapters in the "Notes" sections. No claim to originality is made in case there is a missing reference.

I would like to thank Sara Maad Sasane for going through the entire manuscript, pointing out typos and mistakes, and offering insightful suggestions and comments. Useful comments from Norman Biggs, Rohit Grover, Lassi Paunonen, Rudolf Rupp, Konrad Swanepoel and Victor Ufnarovski are also gratefully acknowledged. Finally, it is a pleasure to thank the staff at World Scientific; in particular, Rochelle Kronzek (Executive editor) for her prompt help and support, and Eng Huay Chionh (Editor) for overseeing the production of the book, and for valuable copy editorial comments which improved the quality of the book.

Amol Sasane
Lund, 2015

Contents

PLAIN PLANE GEOMETRY

Chapter 1

Geometric figures

In this chapter, we will introduce some of the basic geometric characters in our geometric "play" to follow in this book. In some sense, this is the most boring chapter of the book, as it does not have much "action". The real fun in geometry begins when things start interacting. So a certain amount of patience is required. But this is true when learning any new game. One needs to know how the chess pieces move before one can start playing the game, and that part of learning the rules is not very exciting. It is the same with geometry.

We also mention that we will appeal to our intuitive notions of plane geometry, but we will try to be efficient in this. The aim is to start with evident basic truths and rules, and set up new derived rules from these, eventually leading to a body of results and techniques which aren't currently evident. We will see what we mean by this along the way.

1.1 Points, lines, rays, line segments and length

All our action takes place in the *plane*, and the picture we have in mind is that of an infinite flat perfect surface in space, which has no height.

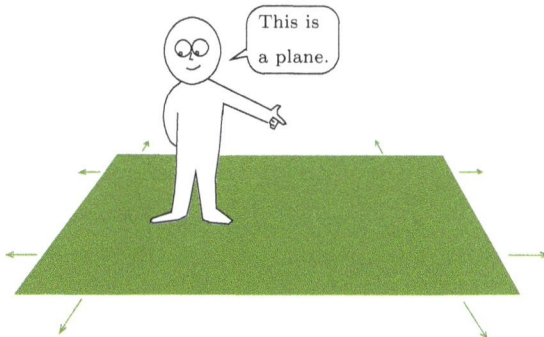

Thus we can talk about objects in the plane which are flat, and have no height, but do possess breadth and length. We usually won't bother depicting the plane in our pictures, and it is assumed that the surface of the page of this book is taken as the plane for our geometrical pictures.

A *point* in the plane is something which has no length or breadth. Imagine making a tiny "dot" on a piece of paper with a pencil. The smaller the dot, the closer it is to representing the geometrical entity of a point in the plane. We will denote points typically by capital letters such as A, P, X, etc. Ideally a point has no "dimensions", and so it is challenging to depict it. Often we will get around this by showing it as a tiny circle.

○ a point

A *line* is something that has no breadth. We imagine our lines to extend on both sides indefinitely, and usually one denotes this by drawing arrows on both sides. Lines will be denoted by small letters such as ℓ, ℓ', ℓ_1 etc.

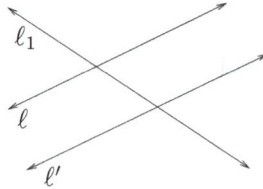

If distinct lines ℓ_1, ℓ_2 in the plane do not intersect, then they are said to be *parallel*, and we write $\ell_1 \parallel \ell_2$. (If $\ell_1 = \ell_2$, then the two lines are coincident, and strictly speaking, we will consider them parallel too[1], although our default stance will be to consider parallel lines as distinct, unless stated otherwise.)

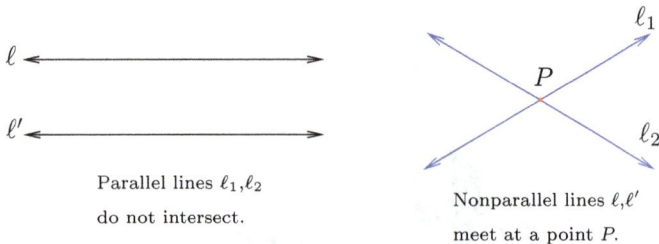

Parallel lines ℓ_1, ℓ_2
do not intersect.

Nonparallel lines ℓ, ℓ'
meet at a point P.

---

[1]So that the relation of "being parallel" becomes an "equivalence relation"; see Exercise 1.10.

Three or more lines in the plane needn't have a point in common, but if they do, then we call the set of lines *concurrent*.

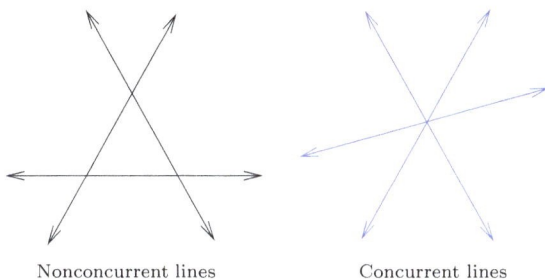

Nonconcurrent lines   Concurrent lines

Infinitely many lines pass through any point in the plane, but given two distinct points in the plane, there is a unique line which passes through them. A line passing through distinct points A, B in the plane will be denoted by $\overleftrightarrow{AB}$.

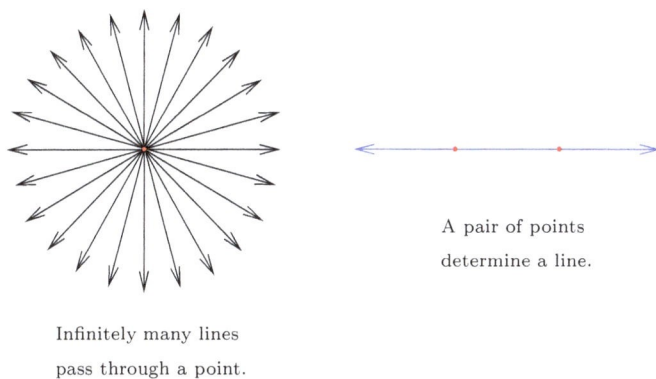

A pair of points
determine a line.

Infinitely many lines
pass through a point.

Three or more points in the plane needn't all lie on a line, but when they do, the set of points are said to be *collinear*.

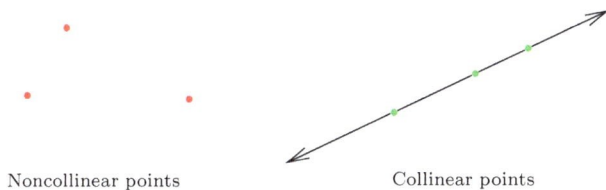

Noncollinear points   Collinear points

A *ray* is part of a line which extends indefinitely in only one direction. If

O is the starting point of the ray, and P is another point on the ray, then we denote the ray by $\overrightarrow{OP}$.

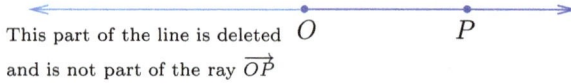

This part of the line is deleted O $\qquad$ P
and is not part of the ray $\overrightarrow{OP}$

A *line segment* is the part of line between two points. We denote the line segment with endpoints A and B by AB or (just as well) by BA. It has finite length, which is measured as a multiple of a fixed, but arbitrarily chosen, "unit length". Ideally one needs different notation to distinguish between the line segment (a geometrical object) and its length (a number), but one just uses the same notation, and the context tells us what is meant. Line segments with equal lengths are usually denoted by putting small identical marks through them.

In the picture below, if one decides that line segment AB is of unit length, then the line segment CD (which can be covered exactly by placing AB two times along its length) has length 2, that is, $CD = 2$. The point E is midway between C and D. Similarly, $MN = \frac{3}{2}$.

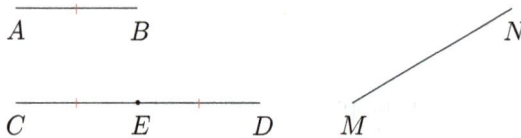

Later on, we will learn about Pythagoras's Theorem, which says that "in any right angled triangle, the sum of the squares of the two sides which include the right angle[2] equals the square of the side opposite the right angle". In the following picture, if $PR = RQ = 1$ (unit length), then the length of PQ must satisfy $PQ^2 = PR^2 + RQ^2 = 1^2 + 1^2 = 2$, and so $PQ = \sqrt{2}$.

---

[2]See the following section for the definition of a right angle.

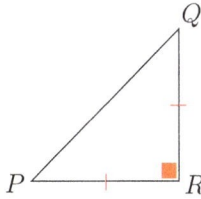

A point M on a line segment AB which divides the line segment into two line segments of equal lengths, (that is, $AM = MB$) is called the *midpoint* of the line segment AB.

1.2 Angles and the degree measure

If two rays share their starting point, then they divide the plane into two parts, one of which is a "wedge". This wedge is referred to as an *angle*. The (common) starting point of the two rays forming the angle is referred to as the *vertex* of the angle, and the two rays are called the *arms* of the angle. The set of points in the wedge between the two rays is called the *interior* of the angle, while the set of points outside the wedge is called the *exterior* of the angle.

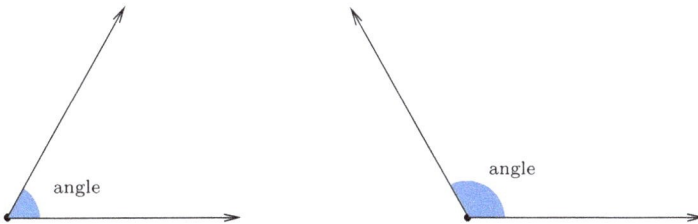

We will denote an angle by a symbol such as $\angle A$, when the vertex is the point A (leftmost picture below), or if there are more than one angle sharing a vertex, then we can specify the angle by specifying points on the arms too, such as $\angle QPR$, as shown in the middle figure. Sometimes, it might be convenient to number the angles, in which case we may simply write $\angle 1, \angle 2$, etc.

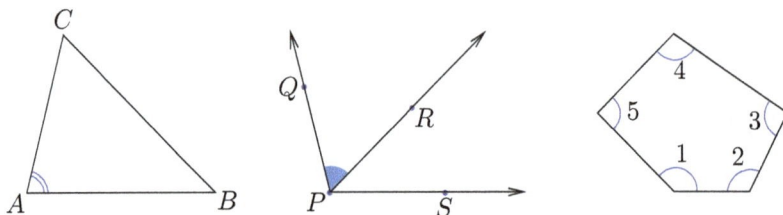

One can measure how big each angle is in the following manner. First consider the case when the two rays emanate from their starting points in opposite directions (so that they together describe a line). Then the two angles are equal, and we say their measure is *two right angles*, or 180 *degrees*, or 180°.

180° or
two right angles

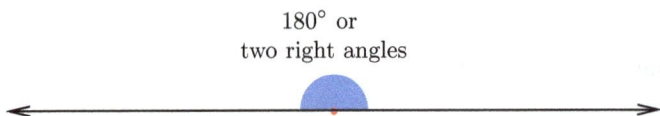

What is a degree? The angle made by opposing rays is divided into 180 equal parts, and each part is then equal to one degree, written 1°. The division into 180 equal parts seems strange, but there are several possible historical explanations of this. We won't discuss this here[3]. The picture below shows a 1° angle.

If an angle of measure 1° is subdivided into 60 equal parts, then each resulting angle is said to have a measure of 1 *minute*, denoted by 1′. Similarly the division into 60 equal parts of an angle of measure 1′ results in each subdivision having an angle measure of 1 *second*, denoted by 1″, and so on.

A 90° angle is the one obtained when the 180° angle is divided into two equal parts, and then the two rays are said to form a *right angle*, or are said to be *perpendicular* to each other. We will usually denote a right angle with the corner of a square as shown in the following picture.

---

[3]We briefly mention that it is believed that having 360° correspond to a full rotation might be related to the fact that some ancient calendars like the Persian calendar divided the year into 360 days, and this in turn may have arisen from the Sumerian use of the "sexagesimal" numeral system with 60 as its base.

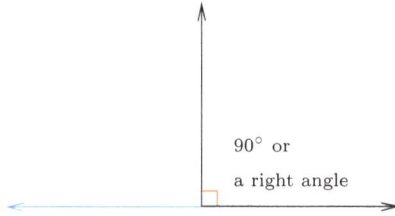

90° or
a right angle

We call an angle *acute* if its measure is less than 90°, and *obtuse* if its measure is more than 90°, but less than 180°.

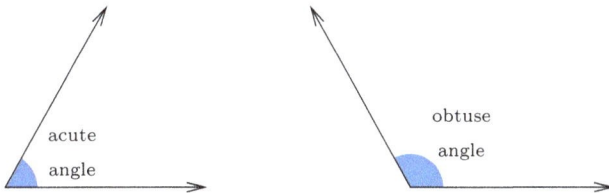

acute
angle

obtuse
angle

If two angles $\angle AOP$ and $\angle POB$ share their vertex O and the common arm $\overrightarrow{OP}$, then the measure of the angle $\angle AOB$ is the sum of the measures of $\angle AOP$ and $\angle POB$.

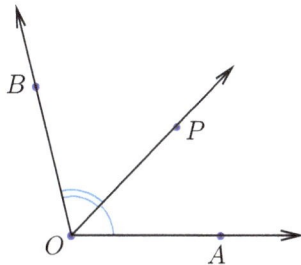

When two lines intersect in a point, then four angles are created.

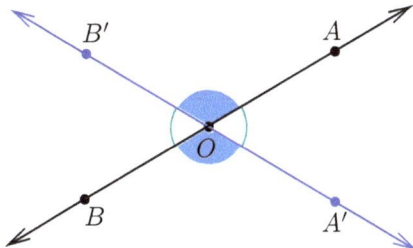

The pairs ($\angle AOA'$, $\angle BOB'$) and ($\angle AOB'$, $\angle BOA'$) are called pairs of *opposite angles*. We claim that the opposite angles are equal. Indeed, as A, O, B lie on a straight line, $\angle BOB' + \angle B'OA = 180°$. Also since the points A', O, B' lie on a straight line, $\angle AOA' + \angle AOB' = 180°$. The above two equations yield $\angle AOA' = \angle BOB'$. Similarly, one can show that $\angle BOA' = \angle AOB'$.

Two lines ℓ_1, ℓ_2 are said to be *perpendicular* to each other, denoted $\ell_1 \perp \ell_2$, if one of the angles created upon intersection angles is 90°. Then it is easy to see that *all* the other angles created are also 90°!

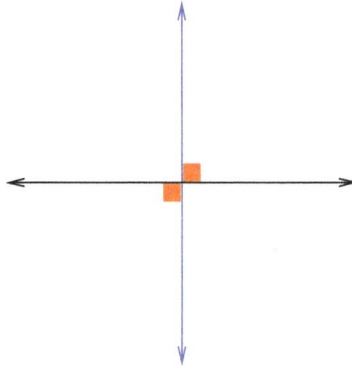

Two angles whose measures add up to 180° are called *supplementary* angles (or supplementary to each other). For example, an angle of measure 120° is supplementary to one having measure 60° since $120 + 60 = 180$.

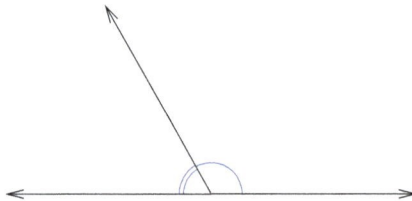

If $\angle AOC$ and $\angle BOC$ are supplementary angles which share the vertex O and the common arm $\overrightarrow{OC}$, then the points B, O, A are collinear. Indeed, we can extend OA to obtain the line $\overleftrightarrow{B'A}$, and then we have

$$\angle BOC + \angle AOC = 180° = \angle B'OC + \angle AOC$$

so that $\angle BOC = \angle B'OC$. Consequently, the ray $\overrightarrow{OB'}$ must coincide with the ray $\overrightarrow{OB}$. As A, O, B' are collinear, and since O, B, B' are collinear, it follows that A, O, B are collinear.

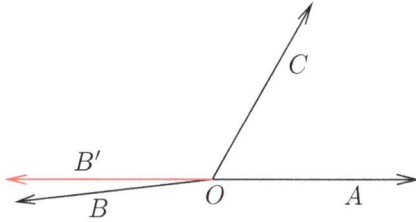

Two angles whose sum of measures add up to 90° are said to be *complementary* angles (or complementary to each other). For example, two angles of measures 60° and 30° form a pair of complementary angles.

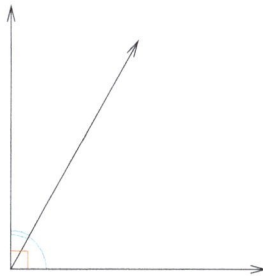

Note that each angle in a pair of complementary distinct angles is acute, while in a pair of supplementary distinct angles, one of the angles is acute and the other is obtuse.

A ray in the interior of an angle which divides the angle into two angles of equal measure is called its *angle bisector*.

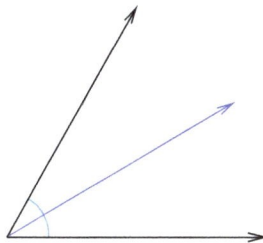

Exercise 1.1. The measure of an angle is 20° more than its complement. What is the measure of the angle?

Exercise 1.2. The measure of an angle is thrice the measure of its supplement. What is the measure of the angle?

Exercise 1.3. If ℓ is a line in the picture below, then what is x?

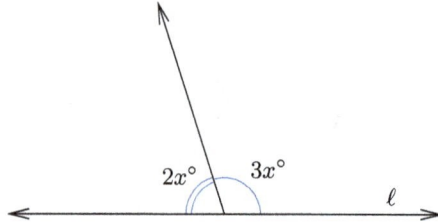

Exercise 1.4. Rays $\overrightarrow{OA}$, $\overrightarrow{OB}$, $\overrightarrow{OC}$, $\overrightarrow{OD}$, $\overrightarrow{OE}$ meet at O as shown in the picture below. Show that $\angle 1 + \angle 2 + \angle 3 + \angle 4 + \angle 5 = 360°$.

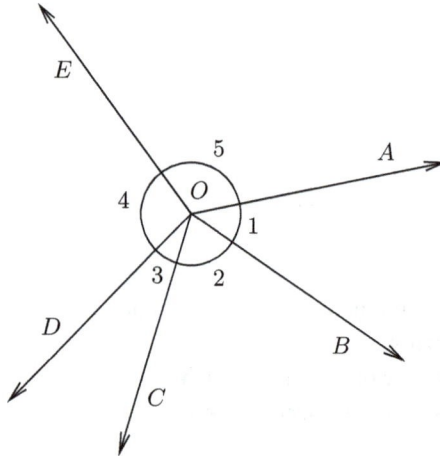

Exercise 1.5. Consider the line $\overleftrightarrow{AB}$ shown below. If $\overrightarrow{OD}$ bisects $\angle COA$ and $\overrightarrow{OE}$ bisects $\angle BOC$, then show that $\overrightarrow{OE} \perp \overrightarrow{OD}$.

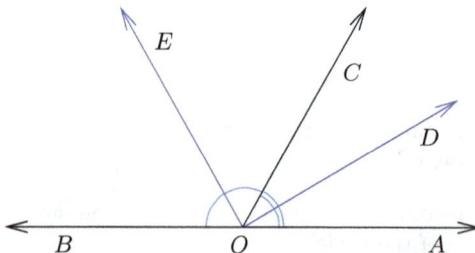

Exercise 1.6. If three concurrent lines intersect as shown in the leftmost picture below, then determine the value of y.

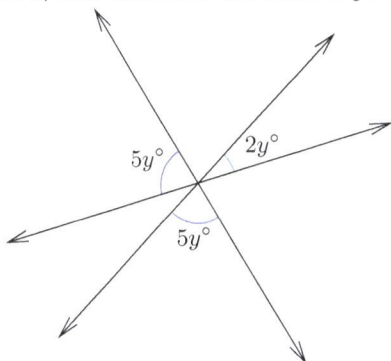

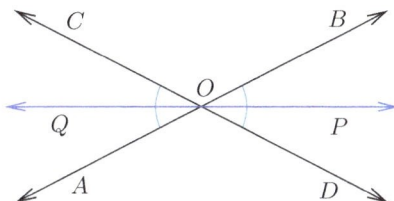

Exercise 1.7. In the rightmost picture above, the two lines $\overleftrightarrow{AB}$ and $\overleftrightarrow{CD}$ intersect at O. $\overrightarrow{OP}$ and $\overrightarrow{OQ}$ are the angle bisectors of $\angle DOB$ and $\angle AOC$ respectively. Show that P, O, Q are collinear.

1.3   The Parallel Postulate

We will accept the following as a rule in planar geometry, and we will call it the "Parallel Postulate"[4]:

The Parallel Postulate: Given a line ℓ and a point P not lying on ℓ, there is a *unique* line $L_{P,\ell}$ which passes through P and is parallel to ℓ.

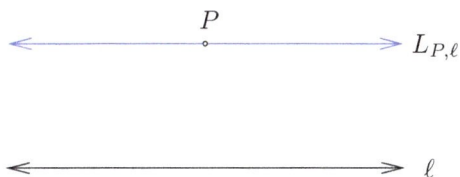

---

[4]Although we call this the Parallel Postulate, the term "Parallel Postulate" usually refers to a different mathematical statement, which is Euclid's fifth postulate in his treatise *Elements*. The version we have written here is called *Playfair's Axiom* (after the Scottish mathematician John Playfair). However, together with the other of Euclid's axioms, Euclid's fifth postulate and Playfair's Axiom can be shown to be equivalent. We have chosen the slightly inaccurate terminology of calling Playfair's Axiom the Parallel Postulate because it helps us remember the statement.

Thus if we consider any other line passing through P which is different from $L_{P,\ell}$, then it must be the case that this other line is not parallel to ℓ.

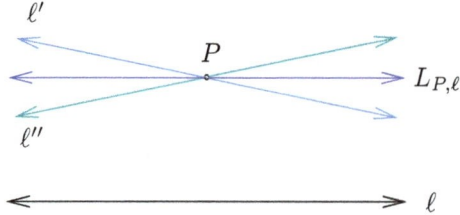

It follows that two distinct intersecting lines cannot both be parallel to the same line.

Exercise 1.8. If ℓ, m, n are lines such that $m \parallel n$ and ℓ intersects m, then show that ℓ intersects n too.

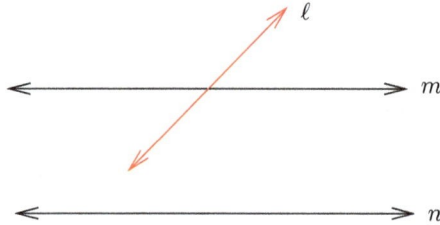

Exercise 1.9. Let A be a point outside the line ℓ, and suppose that $\overleftrightarrow{AB}$ and $\overleftrightarrow{AC}$ are parallel to ℓ. Prove that A, B, C are collinear.

Exercise 1.10. A *relation* R on a set S is a subset of $S \times S := \{(a,b) : a, b \in S\}$. If $(a,b) \in R$, then we write aRb. For example, if we take S to be the set of all human beings, then

$$R_{\text{sibling}} := \{(a,b) \in S \times S : a, b \text{ have the same biological parents}\}$$

is a relation. As another example, we can take the set $S = \mathbb{Z}$, the set of all integers, and $R_{\text{mod } 2} = \{(m,n) \in \mathbb{Z} \times \mathbb{Z} : m - n \text{ is divisible by } 2\}$.

A relation R on a set S is called an *equivalence relation* if it satisfies the following properties:

(ER1) R is *reflexive*, that is, for all $a \in S$, aRa.

(ER2) R is *symmetric*, that is, if aRb, then bRa.

(ER3) R is *transitive*, that is, if aRb and bRc, then aRc.

For example, in our example above, where $S = \{$all human beings$\}$, R_{sibling} can easily be checked to be an equivalence relation[5]. Similarly $R_{\text{mod }2}$ is an equivalence relation on $\mathbb{Z}$.

Why are equivalence relations useful? It helps "partition" the set into "equivalence" classes, and helps to break down the big set into smaller subsets, such that all the elements in each subset are related to each other, and hence "equivalent" in some way. For example, R_{sibling} enables one to partition the set of human beings into equivalence classes consisting of groups of brothers/sisters. On the other hand, $R_{\text{mod }2}$ partitions $\mathbb{Z}$ into the sets $\{$even integers$\}$ and $\{$odd integers$\}$. If R is an equivalence relation of a set S, then we define the *equivalence class of* a, denoted by $[a]$, by

$$[a] = \{b \in S : aRb\}.$$

Given any $a, b \in S$, either $[a] = [b]$ or $[a] \cap [b] = \emptyset$. Indeed, let $[a] \cap [b] \neq \emptyset$. Suppose that $c \in [a] \cap [b]$, that is, aRc and bRc. By symmetry, cRb. As aRc and cRb, by transitivity, we obtain aRb, and again by symmetry, bRa. If $d \in [a]$, then aRd. As bRa and aRd, by transitivity, bRd. So $d \in [b]$ too. So we have shown that $[a] \subset [b]$. In the same manner, one can show that $[b] \subset [a]$ as well. So $[a] = [b]$. We have

$$S = \bigcup_{a \in S} [a],$$

and so every element of S is in *some* equivalence class, and moreover, as any two distinct equivalence classes don't overlap at all, it follows that S is partitioned into equivalence classes by R, as shown in the schematic picture below.

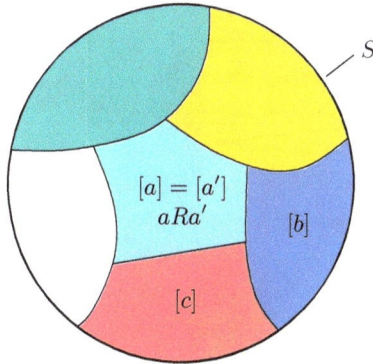

So the idea is that an equivalence relation is really an "attention focusing device", where we have chosen to ignore other distinguishing features of objects which are related, and have put them together in an equivalence class. So an equivalence relation gives one a "pair of glasses" though which we "clump together" things which are essentially the same (equivalent under the relation) and see them as one object! For example, if our set is the collection of children in a school bus and we

---

[5] Here we accept that a person is one's own sibling.

consider the equivalence relation R_1 of "having the same sex", then through these glasses, we see only two equivalence classes: boys and girls. On the other hand, if we consider the equivalence relation R_2 of "having the same age", then through these glasses, we see groups of children sorted by age. We have in fact been using equivalence classes since elementary school: indeed recall that we consider all fractions

$$\frac{1}{2} = \frac{2}{4} = \frac{-399}{-798} = \cdots$$

as "equivalent", describing the same fraction. This can be formalized: a rational number is in fact an equivalence class of $\mathbb{Z} \times (\mathbb{Z} \setminus \{0\})$ (pairs of integers[6]) where two integer pairs (a, b) and (m, n) are related[7] if $a \cdot n = b \cdot m$.

Show that $\parallel$ is an equivalence relation on the set of all lines in a plane.

1.4 The Corresponding Angles Axiom

Suppose that we have two distinct lines ℓ_a, ℓ_b and a third line ℓ which intersects these at two points. The line ℓ is called a *transversal*. We obtain eight angles at the points of intersection of transversal with the lines ℓ_a and ℓ_b. We have labelled these angles $\angle 1_a, \angle 2_a, \angle 3_a, \angle 4_a$ and $\angle 1_b, \angle 2_b, \angle 3_b, \angle 4_b$ as shown in the picture below.

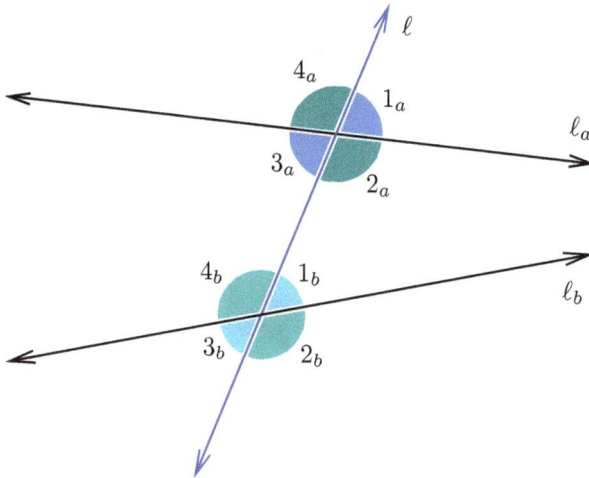

---

[6] In mathematics, it is standard to denote the set of integers $\cdots, -3, -2, -1, 0, 1, 2, 3, \cdots$ by $\mathbb{Z}$, arising from the German word "**zählen**", meaning "count".

[7] The reader might wish to verify that this relation is indeed an equivalence relation.

It will be convenient to give certain angle pairs special names, in order to simplify the discussion in the rest of this book. To this end, we call the pairs of angles

$$(\angle 1_a, \angle 1_b), \quad (\angle 2_a, \angle 2_b), \quad (\angle 3_a, \angle 3_b), \quad (\angle 4_a, \angle 4_b)$$

as pairs of *corresponding angles*. The angles $\angle 2_a, \angle 3_a, \angle 1_b, \angle 4_b$ lie in the region between the given lines ℓ_a and ℓ_b, and can be thought of as "interior angles". With this in mind, we call the pairs of angles $(\angle 2_a, \angle 4_b)$ and $(\angle 3_a, \angle 1_b)$ as *alternate interior angles* since they lie on opposite sides of the transversal ℓ. On the other hand, the pairs $(\angle 2_a, \angle 1_b)$ and $(\angle 3_a, \angle 4_b)$, which lie on the same side of the transversal are called *consecutive interior angles*.

We will accept the following as a rule in planar geometry:

Corresponding Angles Axiom:
If a transversal intersects two parallel lines, then the two angles in each pair of the corresponding angles are equal.

Conversely, if a transversal intersects two lines, making the two angles in a pair of corresponding angles equal, then the lines are parallel.

The following consequences of the Corresponding Angles Axiom are easily seen:

(1) If a transversal intersects two parallel lines, then the angles in each pair of alternate interior angles are equal.

(2) Conversely if a transversal intersects two lines such that the angles in a pair of alternate interior angles are equal, then the two lines are parallel.

(3) If a transversal intersects two parallel lines, then the angles in each pair of consecutive interior angles are supplementary.

(4) Conversely, if a transversal intersects two lines such that the angles in a pair of consecutive interior angles are supplementary, then the two lines are parallel.

For example, let us show statement (4) above. Referring to the previous figure, let us suppose that $\angle 2_a$ and $\angle 1_b$ are supplementary. Then since angles $\angle 1_a, \angle 2_a$ are also supplementary, it follows that $\angle 1_b$ is equal to $\angle 1_a$. As a pair of corresponding angles, namely $(\angle 1_a, \angle 1_b)$, are equal, we conclude by the Corresponding Angles Axiom that $\ell_a \parallel \ell_b$.

Exercise 1.11. In the picture on the left, the two mirrors are parallel to each other. Show that the incident ray AB is parallel to the reflected ray CD.

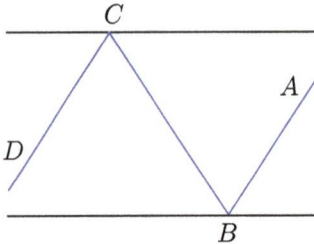

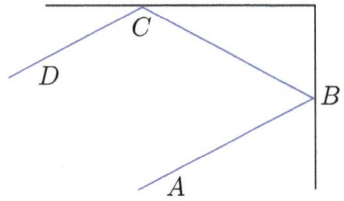

Exercise 1.12. In the picture on the right, the two mirrors are perpendicular to each other. Show that the incident ray AB is parallel to the reflected ray CD.

1.5 Polygons

Besides the elementary lines and points in planar geometry, the other two main objects are polygons and circles. In this section we will describe what we mean by polygons.

The word polygon is derived from Greek, where "poly" means many, and "gonia" means corner/angle. Roughly speaking, it is a geometrical figure made up of a finite number of line segments put end to end in a cyclic manner in order to form a closed figure. Right now, let us learn to recognize a polygon by looking at a couple of pictures. Here are examples of polygons.

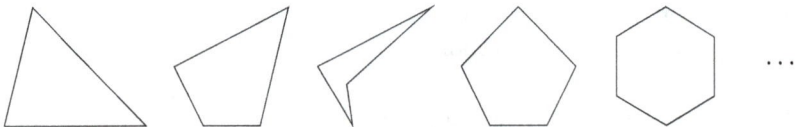

Suppose that $n \geq 3$ distinct points $P_1, \cdots, P_n$ are such that no two line segments among the n line segments

$$P_1 P_2, \quad P_2 P_3, \quad \cdots \quad , \quad P_{n-1} P_n, \quad P_n P_1$$

intersect each other except at their endpoints, and none of the triples of points

$$(P_1, P_2, P_3), (P_2, P_3, P_4), \cdots$$
$$\cdots , (P_{n-2}, P_{n-1}, P_n), (P_{n-1}, P_n, P_1), (P_n, P_1, P_2)$$

are collinear. Then the union of these line segments is called the *polygon* $P_1 \cdots P_n$. The line segments P_1P_2, P_2P_3, $\cdots$, $P_{n-1}P_n$, P_nP_1 are called the *sides* of the polygon, and the points $P_1, \cdots, P_n$ are called the *vertices* of the polygon. The angle determined by two sides meeting at a vertex is called an *angle* of the polygon. The line segments which are obtained by joining the vertices of a polygon, and which are not any of the sides, are called the *diagonals* of the polygon. For example, in the following 5 sided polygon $ABCDE$, we have the 5 possible diagonals AC, AD, BD, BE, CE.

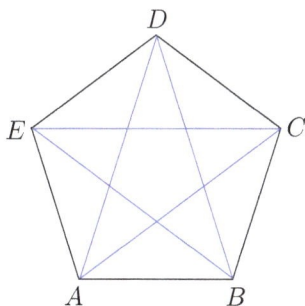

If all the sides and all the angles of a polygon are equal, then we call the polygon a *regular polygon*. In the pictures of polygons at the beginning of this section, the last figure shows a regular polygon with six sides.

Exercise 1.13. How many diagonals does a polygon with n sides have? Check that your formula in terms of n gives the correct answer (5) when $n = 5$.

For small values of n, one gives special names to polygons. For example, when $n = 5$, we call the polygon with 5 sides a *pentagon*, and when $n = 6$, we call the polygon with 6 sides a *hexagon*. In the pictures of polygons shown above, the last two figures are a pentagon and a hexagon. But in this book, we will mostly be concerned with the simplest of polygons, namely triangles and quadrilaterals, which are polygons when $n = 3$, respectively $n = 4$.

If $n = 3$, the polygon is called a *triangle*. Thus a triangle has three sides and three vertices. The name makes sense, since at each vertex, the sides touching at that vertex describes an angle, and thus a triangle is a geometrical figure with three angles: tri + angle = triangle. A triangle with vertices A, B, C will be denoted by $\triangle ABC$, and its angles will be denoted by $\angle A, \angle B, \angle C$ when there is no risk of confusion, and if there might be two triangles sharing a vertex we will be more specific, and use for example the notation $\angle BAC$ to refer to the angle at the vertex A of the triangle $\triangle ABC$.

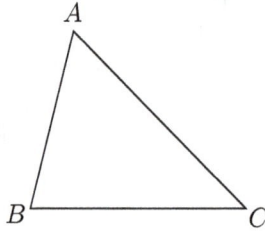

Depending on the lengths of the sides of a triangle, we can classify triangles as scalene/isosceles/equilateral as follows.

If some two sides of a triangle are equal, then the triangle is called an *isosceles triangle*. Etymology: in Greek, iso=same, skeles=leg.

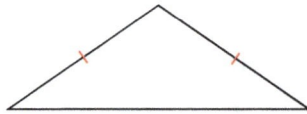

If *all* the sides of a triangle have the same length, then it is said to be *equilateral*.

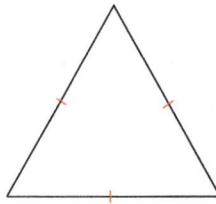

A triangle in which no two sides have equal lengths is called a *scalene triangle*.

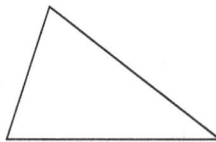

Based on the comparison of the angles of a triangle with the right angle, we can classify triangles into acute/right/obtuse angled triangles as follows.

A triangle which has all angles of measure less than 90° is called an *acute angled triangle*.

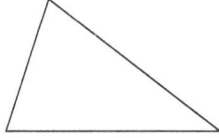

A triangle which has an angle with measure larger than 90° is called an *obtuse angled triangle*.

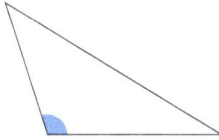

A triangle which has an angle with measure equal to 90° is called a *right angled triangle*.

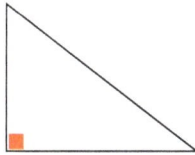

Note that a triangle divides the plane into two regions: the part inside the triangle is called the *interior* of the triangle, and the part outside is called the *exterior* of the triangle.

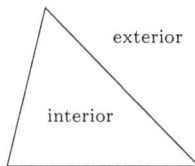

The interior of the triangle has the following property: the line segment formed by joining any two points in the interior of the triangle again lies in the interior of the triangle. In general a region in the plane is said to be *convex* if whenever we take any two points in the region, the line segment joining the two points is again contained in the region. See the following picture.

Exercise 1.14. Given any triangle, does there always exist an "inscribed" equilateral triangle in it, that is, is there an equilateral triangle all of whose vertices lie on the three sides of the original triangle? If so, how many inscribed equilateral triangles are there?

A polygon with four sides and four vertices is called a *quadrilateral*. When specifying a quadrilateral, we will order the vertices in a manner so that adjacent vertices do correspond to a side in the quadrilateral. So $ABCD$, $BCDA$, $CDAB$, $ADCB$ all refer to the quadrilateral shown below with the vertices A, B, C, D, while $ACBD$ doesn't (because otherwise with our notation, AC would be a side in the quadrilateral, but it isn't).

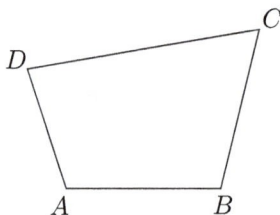

We will not classify quadrilaterals right now, since we will study them in detail in Chapter 3. But we do mention one pertinent point. Just as with triangles, a quadrilateral too divides the plane into two regions: the set of points inside the quadrilateral is called its *interior*, and the set of points outside the quadrilateral is called its *exterior*. However, unlike triangles whose interior is always a convex region, in the case of the interior of a quadrilateral, the interior may not always be convex; see the picture on the right in the following figure.

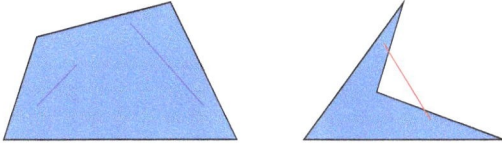

Quadrilaterals whose interiors are convex are called *convex quadrilaterals*. In this book, we will always work with convex quadrilaterals, unless otherwise stated. Some of the theorems we state may also hold for nonconvex quadrilaterals, but we won't make a fuss about stating them in such generality.

Exercise 1.15. Is the interior of an acute/obtuse angle convex? What about its exterior?

Exercise 1.16. Count the number of triangles in the picture shown below.

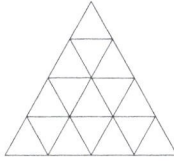

Sum of angles in a triangle

Theorem 1.1. *The sum of the three angles in a triangle is* $180°$.

Proof. Consider a triangle $\triangle ABC$, and extend its base BC to a point D. Draw a ray $\overrightarrow{CE}$ starting from the point C, lying in the interior of $\angle ACD$, which is parallel to the side AB.

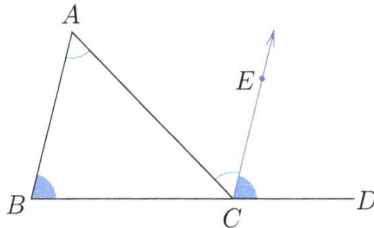

Then the alternate angles $\angle A$ and $\angle ACE$ are equal. Also, $\angle B = \angle ECD$. So $\angle A + \angle B + \angle C = \angle ACE + \angle ECD + \angle ACB = \angle ACD + \angle ACB = 180°$, where we used the fact that $\angle ACB$ and $\angle ACD$ are supplementary in order to obtain the last equality. $\square$

Corollary 1.1. *The non-right angles in a right angled triangle form a pair of complementary angles.*

Proof. If in $\triangle ABC$, $\angle B = 90°$, then

$$180° = \angle A + \angle B + \angle C = \angle A + 90° + \angle C,$$

and so $\angle A + \angle C = 180° - 90° = 90°$. $\qquad\square$

Corollary 1.2. *The sum of the angles of a quadrilateral is* $360°$.

Proof. Let the quadrilateral be $ABCD$. Join the diagonal AC.

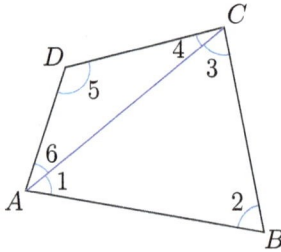

In $\triangle ABC$ and $\triangle ACD$, $\angle 1 + \angle 2 + \angle 3 = 180°$, and $\angle 4 + \angle 5 + \angle 6 = 180°$. Hence

$$\begin{aligned}
\angle A + \angle B + \angle C + \angle D &= (\angle 1 + \angle 6) + \angle 2 + (\angle 3 + \angle 4) + \angle 5 \\
&= (\angle 1 + \angle 2 + \angle 3) + (\angle 4 + \angle 5 + \angle 6) \\
&= 180° + 180° = 360°.
\end{aligned}$$

This completes the proof. $\qquad\square$

Exercise 1.17. If the angles in a triangle are in the ratio $1 : 2 : 3$, find their measures.

Exercise 1.18. Answer the following questions. Can a triangle have

(1) two right angles?
(2) two obtuse angles?
(3) two acute angles?
(4) all angles more than $60°$?
(5) all angles less than $60°$?
(6) all angles equal to $60°$?

Exercise 1.19. The sum of two angles of a triangle is $90°$ and their difference is $30°$. Find all the angles of the triangle.

Exercise 1.20. What is the sum of the angles of a polygon having n sides? Check that your formula gives the right answer ($360°$) when $n = 4$. What is the measure of each angle of a regular polygon having n sides? If a regular polygon with n sides has each angle equal to $120°$, then what is n?

Exercise 1.21. Let $ABCDE$ be a regular pentagon, and suppose that the angle bisector of $\angle A$ meets the side CD at the point M. Show that $AM \perp CD$.

Exercise 1.22. Given a hexagram or a "six-pointed star" formed by the points A, B, C, D, E, F (shown below), show that $\angle A + \angle B + \angle C + \angle D + \angle E + \angle F = 360°$.

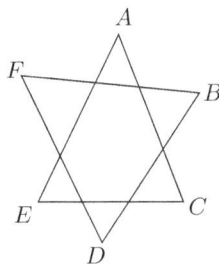

Exercise 1.23. What is the sum of the four angles a, b, c, d in the picture below if it is given that $\ell \parallel \ell'$?

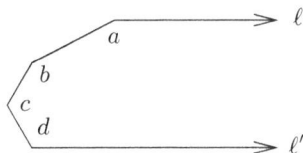

In a triangle, when the sides are produced[8], we get four angles at each vertex. The one opposite to the angle in the triangle is equal to it. There is one more pair of equal and opposite angles obtained, and each of these angles is referred to as an exterior angle of the triangle (at the chosen vertex). Thus in any triangle, there are essentially 6 exterior angles, namely 2 equal exterior angles at each vertex. Thus there are three exterior angles of interest in a triangle. Given an exterior angle, we can consider the opposite two interior angles of a triangle as shown in the rightmost picture in the following figure.

---

[8]We will often say "produced" instead of "extended/continued linearly", with the latter having the obvious meaning.

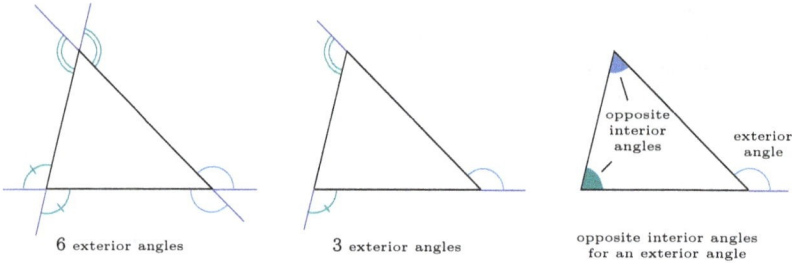

6 exterior angles   3 exterior angles   opposite interior angles for an exterior angle

Corollary 1.3 (Exterior Angle Theorem). *In any triangle, an exterior angle is equal to the sum of the opposite interior angles.*

Proof. Consider the exterior angle in $\triangle ABC$ at the vertex C. Let the interior angles of the triangle be denoted as usual by $\angle A, \angle B, \angle C$. Then $\angle A + \angle B + \angle C = 180°$.

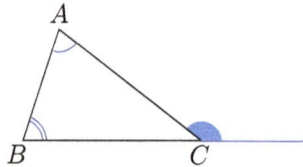

We have

$$\text{exterior angle at } C = 180° - \angle C$$
$$= 180° - (180° - (\angle A + \angle B)) = \angle A + \angle B.$$

$\square$

Exercise 1.24. In a triangle $\triangle ABC$, the angle bisectors of the angles $\angle B$ and $\angle C$ meet at O. Prove that $\angle BOC = 90° + \angle A/2$.

Exercise 1.25. In the left picture below, the angle bisectors of a pair of exterior angles at the vertices B, C meet at O as shown. Prove that $\angle BOC = 90° - \angle A/2$.

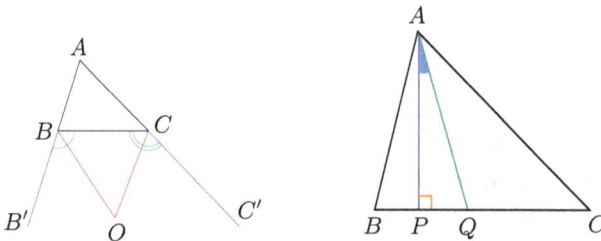

Exercise 1.26. In the right picture above, AQ bisects $\angle BAC$ while $AP \perp BC$. Show that $\angle PAQ = (\angle B - \angle C)/2$.

Exercise 1.27. Given a pentagram or a five-pointed "star" formed by A, B, C, D, E (shown below), show that $\angle A + \angle B + \angle C + \angle D + \angle E = 180°$.

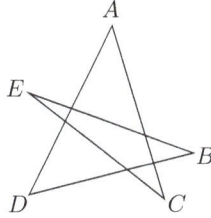

Exercise 1.28. In any acute angled triangle, the altitude dropped from any vertex meets the opposite side at a point in the interior of that side.

Exercise 1.29. If two lines ℓ, ℓ' are met by a transversal such that the corresponding interior angles on one side of the transveral have a sum strictly less than $180°$, then show that ℓ, ℓ' meet at a point on that side of the transversal.

1.6   Circles

Definition 1.1 (Circle). Given a point O in the plane and an $r > 0$, the set of all points in the plane which are at a distance r from O is called the *circle* with *center O* and *radius r*. A circle with center O and radius r will be denoted by $C(O, r)$. See the picture on the left below.

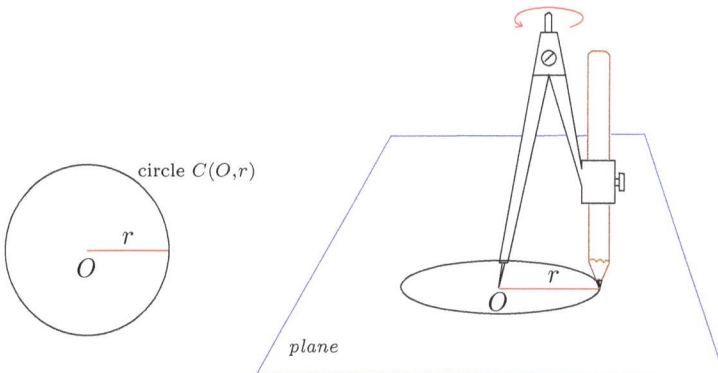

A circle can be drawn with a *compass* as demonstrated in the picture on the right above.

In plane geometry, a *construction* will be an algorithm or a method or a set of instructions to draw a geometrical figure starting from given data, using (usually) an unmarked ruler (straight edge) and a compass. Such recipes will be useful to understand and develop planar geometry, and will serve as useful problems to illustrate how the concepts and theorems we learn can be applied.

Construction 1.1 (Equilateral triangle with a given side length).
Given a line segment of length $a > 0$, construct an equilateral triangle with each side having length a.

Let A, B be the endpoints of the given line segment of length a.

    Step 1. With A as center and radius a, draw circle $\mathcal{C}(A, a)$.

    Step 2. With B as center and radius a, draw circle $\mathcal{C}(B, a)$.

    Step 3. Denote by C any one of the two points of intersection of $\mathcal{C}(A, a)$ and $\mathcal{C}(B, a)$.

    Step 4. Join A to C, and B to C. Then $\triangle ABC$ is an equilateral triangle with each side having length a.

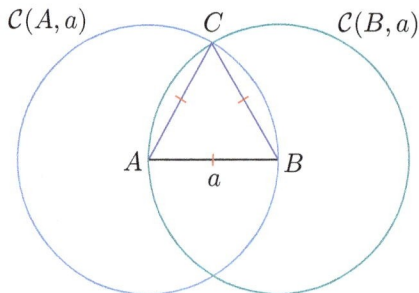

Justification: As $\mathcal{C}(A, a)$ is a circle, and the points B, C lie on it, we have $AB = AC$. Similarly, since $\mathcal{C}(B, a)$ is a circle, and the points A, C lie on it, we also have $AB = BC$. Putting these together, we obtain that $AC = BC = AB = a$, that is, $\triangle ABC$ is an equilateral triangle, with each side of length a. $\diamond$

Exercise 1.30. ($*$)
(Collapsible compass + straight edge ≡ compass + straight edge). As opposed to an (ordinary) compass described above, with which we can "lift" distances from one point to another, a *collapsible compass* is one which allows drawing a circle as before, but we cannot lift distances from one point to another. (A practical way to imagine a collapsible compass is to think of a spring attached between the two "prongs" of a regular compass. This means that while you are drawing a

circle, the friction with the page prevents the compass from collapsing and hence allows you to draw your circle. However, the moment you lift this compass, the friction disappears, causing the spring to contract, and the compass "collapses", preventing retention of the radius faithfully from one point to another.) Euclid wanted to create his constructions using such a collapsible compass. It turns out that one *can* lift distances using a collapsible compass using a construction, and the aim of the exercise is to ask the reader to think of a construction for this.

Given a line segment with end points A, B, a point P in the plane, give a procedure for constructing a line segment PQ having the same length as that of AB, using a collapsible compass and a straight edge alone.

(Once one has this construction, we realize that we don't have to stick to a collapsible compass, but instead we can just as well imagine that we have a regular one at our disposal, since any construction which makes use of a regular compass, can now be done instead with a collapsible compass.)

Notes

Exercises 1.11 and 1.12 are taken from [Vaidya, Rao and Singh (1988)].

Chapter 2

Congruent triangles

Two figures which can be obtained from each other by a rigid motion (rotation, translation or reflection) are considered to be the "same" or *congruent* in planar geometry. Indeed they have the same shape and size. For example, line segments of equal lengths are congruent, angles of equal measure are congruent, circles of the same radius are congruent. In this chapter, we will study congruency of triangles.

Definition 2.1. Two triangles $\triangle ABC$ and $\triangle A'B'C'$ are said to be *congruent* if the corresponding sides are equal and the corresponding angles are equal, that is,

$$AB = A'B', \quad BC = B'C', \quad CA = C'A'$$
$$\angle A = \angle A', \quad \angle B = \angle B', \quad \angle C = \angle C'.$$

We then write $\triangle ABC \simeq \triangle A'B'C'$.

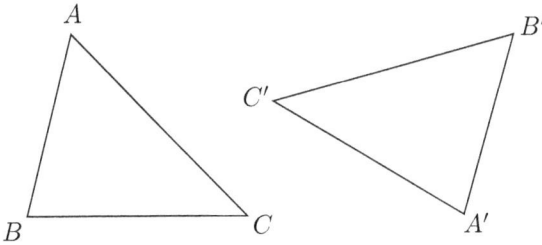

We note that in our notation $\triangle ABC \simeq \triangle A'B'C'$, we have made sure that the vertices correspond to each other in the correct order:

$$A \longleftrightarrow A',$$
$$B \longleftrightarrow B',$$
$$C \longleftrightarrow C'.$$

The goal of this chapter is to first of all establish some simple criteria for the congruency of triangles. Thus rather than checking that *all* corresponding sides and *all* corresponding angles match, it will turn out that one can get away with just doing part of the work. Usually three bits of appropriate data being equal will yield congruency. For example, we will see that if the corresponding sides match, then the triangles will be congruent; this will be called the "SSS Congruency Rule".

Secondly, we will prove additional properties of geometric figures by searching for congruent triangles, and making use of the fact that the corresponding sides and angles are then equal. Since we will need the phrase

"because the corresponding parts of congruent triangles are equal"

very often, we will abbreviate it by

"CPCT".

This will make arguments shorter, since we will all know what we mean when we write CPCT, without laboriously spelling it out fully every time we need to use this.

Exercise 2.1. Show that congruency is an equivalence relation of the set of all triangles in the plane.

2.1   SAS Congruency Rule

The first congruency rule is the SAS Congruency Rule, which we will accept as a self-evident truth in planar geometry. The statement is as follows.

SAS Congruency Rule:
If two sides and the included angle of one triangle are equal to
  two sides and the included angle of another triangle,
then the two triangles are congruent.

Let us see why it makes sense to accept this as an axiom. We observe that knowing two sides, say AB, BC, and the included angle $\angle B$ determines the third side AC automatically. See the picture below. Hence the whole triangle $\triangle ABC$ is determined, that is, we can't envisage two different triangles with the sides AB, BC and the included angle $\angle B$. If SAS Congruency Rule were wrong, then it would violate this natural expectation!

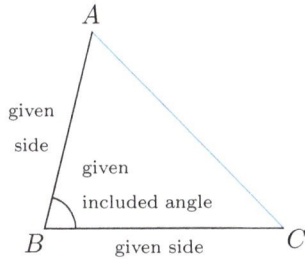

It is important to realize that the fact that the angle is *included* between the given sides is important for drawing the congruency conclusion. If it so happens that two corresponding sides are equal in two triangles and a *nonincluded* angle is equal, then it is not guaranteed that the two triangles are congruent.

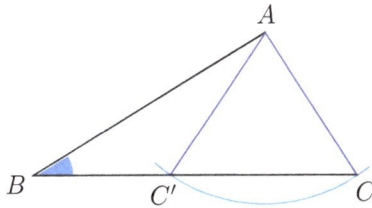

Consider for example the picture above, where the big triangle $\triangle ABC$ (with an obtuse angle at vertex A) is visibly not congruent to the little triangle $\triangle ABC'$, but two corresponding sides and a *nonincluded* angle are equal. (The point C' has been obtained as the intersection point of a circle with center A and radius AC.) Indeed, we have:

$$AB = AB \quad \text{(common side)}$$
$$AC = AC' \quad \text{(construction with the compass)},$$
$$\angle B = \angle B \quad \text{(common angle)}.$$

Later on, we will learn Theorem 2.5, where with extra data preventing the above scenario from happening, we can conclude that triangles with two sides and one nonincluded angle are congruent.

Proposition 2.1. *In any isosceles triangle, the angles opposite the equal sides are equal.*

Proof. Let $\triangle ABC$ be an isosceles triangle with $AB = AC$. Consider the angle bisector of the angle $\angle BAC$, meeting the side BC at the point D.

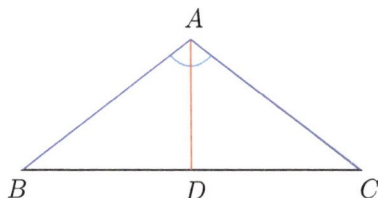

We will show that $\triangle BAD \simeq \triangle CAD$. We have

$$AD = AD \qquad \text{(common side)},$$
$$\angle BAD = \angle CAD \quad \text{(construction)},$$
$$AB = AC \qquad \text{(given)}.$$

So by the SAS Congruency Rule, it follows that $\triangle BAD \simeq \triangle CAD$. Consequently $\angle ABD = \angle ACD$ (CPCT). This completes the proof.

Alternatively, here's a slicker proof: By the SAS Congruency Rule, we have that $\triangle ABC \simeq \triangle ACB$ (because $AB = AC$ and $\angle BAC$ is common to both). Thus $\angle ABC = \angle ACB$ (CPCT). $\qquad\qquad\square$

Proposition 2.2. *In a triangle $\triangle ABC$, if $AB < AC$, then $\angle B > \angle C$.*

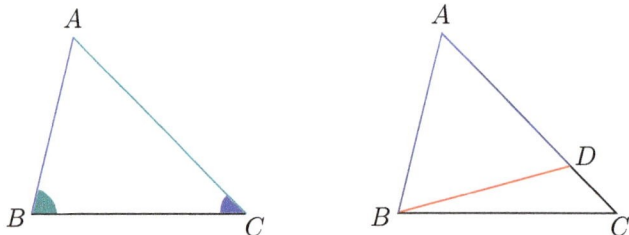

Proof. Let D be a point on AC such that $AD = AB$. Then $\triangle ABD$ is an isosceles triangle. So $\angle ABD = \angle ADB$. Also, by the Exterior Angle Theorem, we have that $\angle ADB = \angle DBC + \angle C$. Consequently

$$\angle B = \angle ABD + \angle DBC > \angle ADB = \angle DBC + \angle C > \angle C.$$

$\qquad\qquad\square$

We will now prove the converse of the above result.

Proposition 2.3. *In a triangle $\triangle ABC$, if $\angle B > \angle C$, then $AB < AC$.*

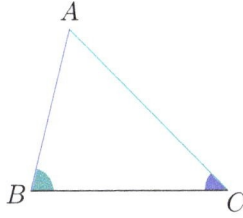

Proof. We have three possible cases.

1° $AB = AC$. Then $\triangle ABC$ is isosceles and $\angle B = \angle C$, a contradiction. So this case is not possible.
2° $AB > AC$. By Proposition 2.2, $\angle C > \angle B$, again a contradiction. So this case is not possible either.
3° Thus the only remaining possible case is that $AB < AC$.

$\square$

Theorem 2.1 (Triangle Inequality). *In any triangle, the sum of the lengths of any two sides is strictly bigger than the remaining side.*

Proof. Let the triangle be $\triangle ABC$ with vertices A, B, C. We will show that $AB + AC > BC$. To this end, let us extend BA to a point D such that $AD = AC$. Then $\triangle ACD$ is isosceles and $\angle ACD = \angle BDC$.

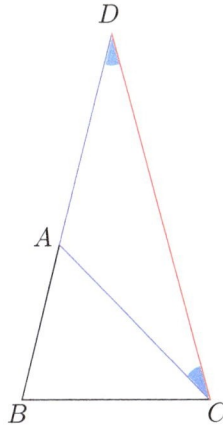

Then $\angle BCD = \angle ACB + \angle ACD > \angle ACD = \angle BDC$. By Proposition 2.3, we have $BD > BC$. But $BD = AB + AD = AB + AC$. Hence we obtain $AB + AC = BD > BC$. $\square$

Exercise 2.2. Show that in a triangle with side lengths a, b, c, the semiperimeter

$$s := \frac{a+b+c}{2}$$

is bigger than each of the side lengths a, b, c. Thus $s > \max\{a, b, c\}$, and so $s - a, s - b, s - c$ are all positive.

Exercise 2.3. (∗) In any triangle $\triangle ABC$, if P is a point in its interior, then show that

$$s < AP + BP + CP < 2s,$$

where $s := \dfrac{AB + BC + CA}{2}$ is the semiperimeter of $\triangle ABC$.

Exercise 2.4. Let a line ℓ be given, along with two points A, B on one side of ℓ. Where is the point P on ℓ which minimizes $PA + PB$?

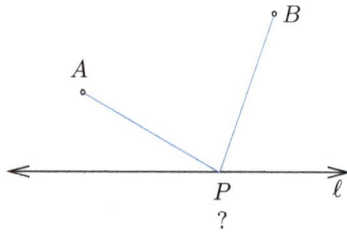

Lemma 2.1 (Dogleg[1] Rule). *If a triangle and a polygon have a side in common and if the triangle lies inside the polygon, then the triangle has a smaller perimeter than the polygon.*

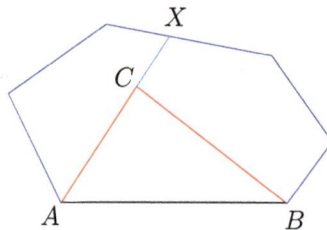

Proof. Let AB be the common side. Extend AC, and let it meet the polygon at a point X. Join X to B. It can be seen by using the Triangle Inequality, that the polygonal path from A to X is bigger than AX. Similarly the polygonal path from X to B is bigger than XB. Thus the polygonal perimeter P satisfies

$$P \geq AB + AX + XB = AB + (CA + CX) + XB$$
$$= AB + CA + (CX + XB)$$
$$\geq AB + AC + CB,$$

as claimed. $\qquad\qquad\qquad\qquad\qquad\qquad\qquad\qquad\qquad\qquad$ □

2.2 SSS Congruency Rule

Theorem 2.2 (SSS Congruency Rule). *If the three sides of one triangle are equal to the corresponding three sides of another triangle, then the two triangles are congruent.*

Proof. Let the triangles with equal corresponding sides be $\triangle ABC$ and $\triangle A'B'C'$. Draw a ray $\overrightarrow{B'A''}$ with starting point B' such that we have $\angle A''B'C' = \angle B$, and also such that the points A'' and A' lie on opposite sides of $B'C'$. Moreover, let the point A'' on the ray be such that we have $B'A'' = AB$. Finally join A'' to C', and also A'' to A'.

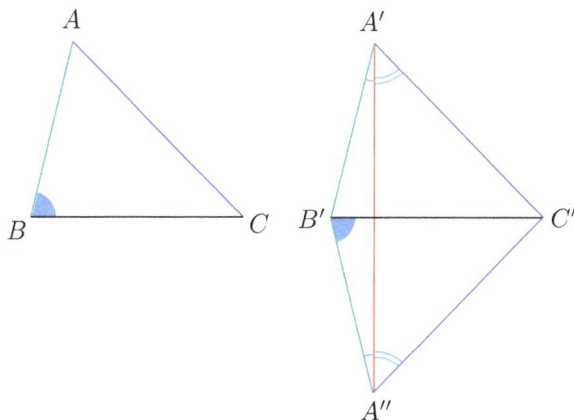

Consider $\triangle ABC$ and $\triangle A''B'C'$. We have that $BC = B'C'$, $AB = A''B'$, and also $\angle B = \angle A''B'C'$. By the SAS Congruency Rule, we conclude that $\triangle ABC \simeq \triangle A''B'C'$. Hence we obtain $\angle A = \angle B'A''C'$ and $AC = A''C'$ (CPCT).

Since $A'C'(= AC) = A''C'$, $\triangle A'C'A''$ is isosceles, and so we have $\angle A''A'C' = \angle A'A''C'$. Also, as $A''B'(= AB) = A'B'$, $\triangle A'B'A''$ is isosceles, and so $\angle B'A'A'' = \angle B'A''A'$. Thus

$$\angle A = \angle B'A''C' = \angle B'A''A' + \angle A'A''C' = \angle B'A'A'' + \angle A''A'C' = \angle A'.$$

Now in the two original triangles $\triangle ABC$ and $\triangle A'B'C'$, besides the equality of the sides $AB = A'B'$ and $AC = A'C'$, also the included angles satisfy $\angle A = \angle A'$. Consequently, by the SAS Congruency Rule, $\triangle ABC \simeq \triangle A'B'C'$. □

Exercise 2.5. (AAA Congruency Rule?) If the three angles of one triangle are equal to the corresponding three angles of another triangle, then must the two triangles be congruent?

Construction 2.1 (Triangle with given side lengths).
Given the lengths of the sides of a triangle, construct the triangle.

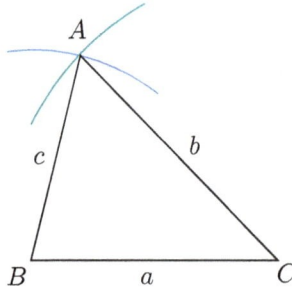

Let the sides have lengths a, b, c.

Step 1. Draw a segment BC of length a.
Step 2. With B as center and radius c, draw the circle $\mathcal{C}(B, c)$.
        With C as center and radius b, draw the circle $\mathcal{C}(C, b)$.
        Let A be any one of the two points of intersection of $\mathcal{C}(B, c)$ and $\mathcal{C}(C, b)$.
Step 3. Join A to B and A to C.

Then $\triangle ABC$ is the triangle with $AB = c$, $BC = a$, $CA = b$.
Justification: The SSS Congruency Rule shows that all such triangles are congruent. ◇

Construction 2.2 (Angle bisector). *Given an angle, construct its angle bisector.*

Before we see this construction, we remark that since we can "construct" 180°, by just taking a point on a line, we can bisect this angle, getting

90°. Similarly we can keep bisecting the resulting angles, getting $45°, 22\frac{1}{2}°$, and so on. Hence the angles $90°, 45°, 22\frac{1}{2}°$, etc., are constructible using a straight edge and compass.

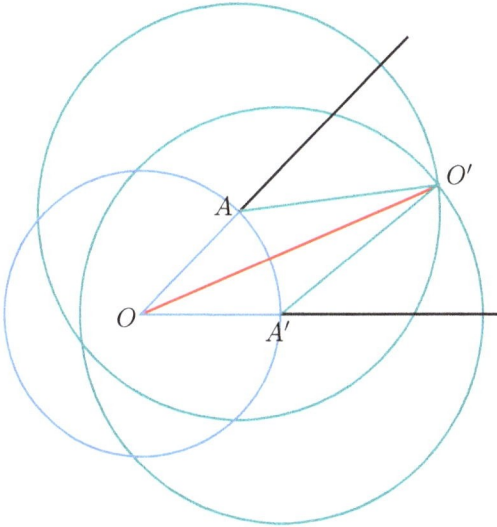

Let the given angle have vertex O.

Step 1. With O as center and with an arbitrary radius $r > 0$, draw a circle $C(O, r)$, and let it intersect the two rays of the angle at the points A, A'.

Step 2. With A as center and a large enough radius r', draw the circle $C(A, r')$.

With A' as center and radius r', draw the circle $C(A', r')$.

Let O' be the point of intersection of $C(A, r')$ and $C(A', r')$, in the interior of the given angle with vertex O.

Step 3. Join O to O'.

The ray $\overrightarrow{OO'}$ bisects the given angle $\angle O$.

Justification: Consider the triangles $\triangle OAO'$ and $\triangle OA'O'$. We have that $OA = OA' = r$ since A, A' lie on $C(O, r)$. Also, $AO' = A'O'$ because O' lies on $C(A, r')$ and on $C(A', r')$. Finally the side OO' is common to both triangles. Hence by the SSS Congruency Rule, $\triangle OAO' \simeq \triangle OA'O'$. Consequently $\angle AOO' = \angle A'OO'$ (CPCT). ◇

Remark 2.1 (Impossibility of angle *trisection*). An ancient problem considered by the Greeks was that of giving a construction method, using only a compass and a straight edge, for the trisection of an arbitrary given angle. The problem resisted all attempts to solve it, and eventually in 1837, it was shown by Pierre Wantzel, using algebraic methods ("Galois Theory"), that it is impossible to give such a procedure! Essentially the trisection of an angle amounts to be able to construct the solution of a certain cubic equation, which is impossible using the given tools. Note, however, that the impossibility of trisecting a *general* angle using a compass and a straight edge does not of course mean that there aren't trisectible angle using these tools: for example, since we know how to draw 45°, we *can* trisect the $3 \cdot 45° = 135°$ angle! The interested reader is referred to [Courant and Robbins (1979)] for a proof of the impossibility of angle trisection.

Construction 2.3 (Reproducing an angle). *Given an angle $\angle O$ and an arbitrary ray $\overrightarrow{O'X}$, construct the angle $\angle YO'X$ such that $\angle YO'X = \angle O$.*

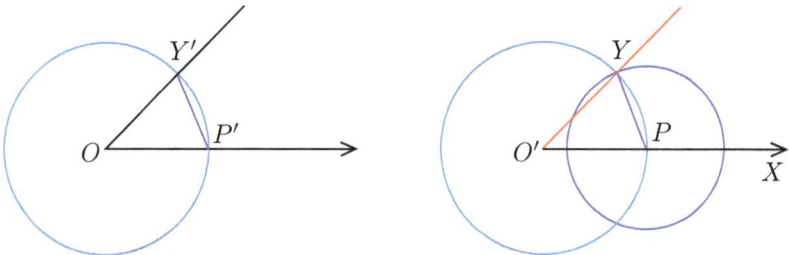

Step 1. With O as center and with an arbitrary radius $r > 0$, draw a circle $C(O, r)$, and let it intersect the two rays of the angle at the points P', Y'.

Step 2. With O' as center and radius r, draw the circle $C(O', r)$, and let it intersect the given ray $\overrightarrow{O'X}$ at the point P.

Step 3. Taking the radius $P'Y'$ and with center P, draw the circle $C(P, P'Y')$, intersecting $C(O', r)$ at two points, any of which can be taken as the point Y.

Then $\angle YO'P = \angle O$.

Justification: In $\triangle OP'Y'$ and $\triangle O'PY$, $OP' = O'P = r = OY' = O'Y$ and $P'Y' = PY$, and so by the SSS Congruency Rule, $\triangle OP'Y' \simeq \triangle O'PY$. Consequently, $\angle O = \angle YO'P$.   ◇

Hence we can double angles, triple angles, and so on. Since we can also bisect angles, given an angle of measure $\alpha°$, we can construct angles of measure

$$\frac{n}{2m}\alpha°,$$

where n, m are arbitrary natural numbers.

Construction 2.4 (Parallel to given line through given point).
Given a line ℓ and a point P outside it, construct the line $L_{\ell,P}$ which passes through P, and is parallel to ℓ.

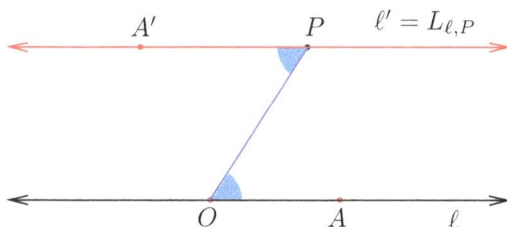

Step 1. Take any point O on ℓ and another point A on ℓ which is on the right of O. Join O to P.

Step 2. Using Construction 2.3, construct the angle $\angle A'PO$ such that $\angle A'PO = \angle POA$ and such that A' and A lie on opposite sides of the line OP.

Then the line ℓ' determined by A' and P is the required line $L_{\ell,P}$.

Justification: $\angle A'PO$ and $\angle POA$ are the alternate interior angles for the lines ℓ' and ℓ with respect to the transversal OP. As these alternate angles are equal, it follows that $\ell' \parallel \ell$.   ◇

Corollary 2.1. *Each angle in an equilateral triangle is equal to $60°$.*

Proof. Let the equilateral triangle be $\triangle ABC$. As all sides have equal lengths, it follows that $\triangle ABC \simeq \triangle BCA \simeq \triangle CAB$. So $\angle A = \angle B = \angle C$ (CPCT). Since the sum of the angles must be $180°$, we conclude that each must be $60°$.   □

Since we know how to construct an equilateral triangle (Construction 1.1), we can now construct a $60°$ angle. Moreover, since we also know how to bisect angles, we can construct $30°, 15°, 7\frac{1}{2}°$, and so on.

Exercise 2.6. Given the length AB, construct a regular hexagon having each side of length AB.

Exercise 2.7. (∗) Given an angle of measure 19°, give a method for dividing it into 19 equal parts using a straight edge and compass.

Exercise 2.8. (∗) (A Euclidean Ramsey Theorem[1]). Given any colouring of the plane by two colours, there exists a *monochromatric* equilateral triangle, that is, an equilateral triangle, all of whose vertices have the same colour.

If AB is a line segment, then a line ℓ which passes through the midpoint M of AB and which is perpendicular to AB is called the *perpendicular bisector* of AB.

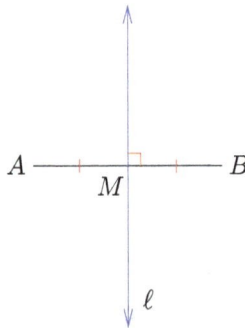

Proposition 2.4 (Perpendicular Bisector Theorem). *Let AB be line segment. A point P in the plane is equidistant from the endpoints of the line segment if and only if P lies on the perpendicular bisector of AB.*

Proof.
(If part): Let P lie on the perpendicular bisector of AB, and let M be the midpoint of AB. Join P to A and B. We'll show $\triangle AMP \simeq \triangle BMP$. We have

$$PM = PM \quad \text{(common side)},$$
$$AM = MB \quad (M \text{ is the midpoint of } AB),$$
$$\angle AMP = \angle BMP = 90°.$$

Thus by the SAS Congruency Rule, it follows that $PA = PB$ (CPCT). Hence P is equidistant from the endpoints A, B of the line segment AB.

---

[1]More generally, in "Euclidean Ramsey Theory", named after the British mathematician Frank Ramsey, one investigates the existence of monochromatic figures possessing a desired set of geometric properties, given colourings of points of the plane with a certain set of paints.

If part   Only if part

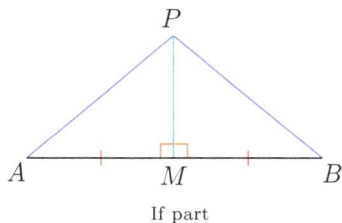

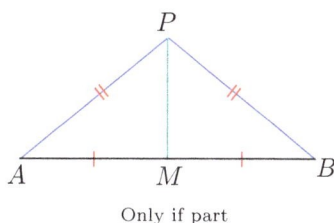

(Only if part): Let P be a point equidistant from A and B. Let M be the midpoint of AB, and join P to M. We claim that $\triangle AMP \simeq \triangle BMP$. We have

$$PM = PM \quad \text{(common side)},$$
$$AM = MB \quad (M \text{ is the midpoint of } AB),$$
$$AP = PB \quad \text{(given)}.$$

Thus by the SSS Congruency Rule, it follows that $\angle PMA = \angle PMB$ (CPCT). But as A, M, B are collinear, $\angle PMA + \angle PMB = 180°$. Hence $2\angle PMB = 180°$, that is, $\angle PMB = 90°$. So PM is perpendicular to AB, and as M is the midpoint of AB, we conclude that the line joining P and M is the perpendicular bisector of AB. Consequently P lies on the perpendicular bisector of AB. $\square$

Construction 2.5 (Bisecting a line segment). *Given a line segment AB, construct its midpoint.*

Given   Wanted

Step 1. Choose an R which is bigger than half the length of AB (for example, we could just take $R = AB$). With A as center and radius R, draw a circular arc. With B as center and radius R, draw a circular arc. Let these two circular arcs meet at the points P_1, P_2.

Step 2. Join $P_1 P_2$, and let it meet AB at P. Then P is the midpoint of AB.

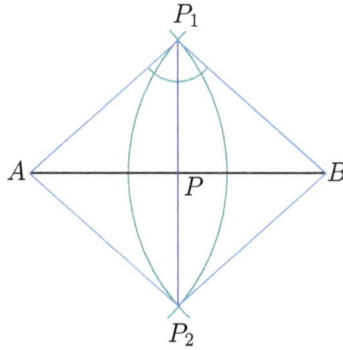

Justification: In triangles $\triangle AP_1P_2$ and $\triangle BP_1P_2$, we have $AP_1 = BP_1$, $AP_2 = BP_2$, $P_1P_2 = P_1P_2$, and so $\triangle AP_1P_2 \simeq \triangle BP_1P_2$ by the SSS Congruency Rule. Hence $\angle AP_1P_2 = \angle BP_1P_2$. But also $AP_1 = BP_1$ and $P_1P = P_1P$, and so by the SAS Congruency Rule, $\triangle AP_1P \simeq \triangle BP_1P$. Consequently $AP = PB$ (CPCT). $\diamond$

Construction 2.6 (Dropping a perpendicular). *If ℓ is a given line and the point O does not lie on ℓ, then construct a perpendicular from O to the line ℓ.*

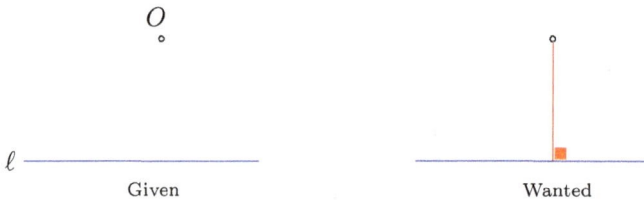

Step 1. Draw a circular arc with center O intersecting ℓ at two distinct points P_1, P_2.

Step 2. Bisect the line segment P_1P_2 to obtain its midpoint P. Then $OP \perp \ell$.

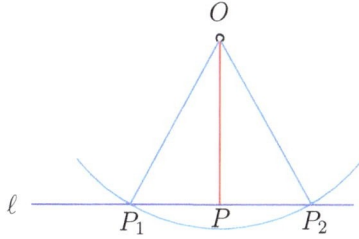

Justification: In $\triangle OPP_1$ and $\triangle OPP_2$, we have

$$OP_1 = OP_2 \quad \text{(both equal to the radius of the circular arc)},$$
$$OP = OP \quad \text{(common side)},$$
$$P_1P = PP_2 \quad \text{(P is the midpoint of P_1P_2)}.$$

By the SSS Congruency Rule, $\triangle OPP_1 \simeq \triangle OPP_2$. Thus $\angle OPP_1 = \angle OPP_2$ (CPCT). As $\angle OPP_1 + \angle OPP_2 = 180°$, we have $\angle OPP_1 = 90° = \angle OPP_2$.
◇

Perpendicular bisectors in a triangle are concurrent

In any triangle, we can construct the perpendicular bisectors of each of the sides of the triangle. We will show that, remarkably, these are concurrent. Their point of concurrency is called the *circumcenter* of the triangle. We note that with O as a center and radius $OA =: R$, we can draw a circle. The points B and C will then lie on this circle too. (Indeed, as O lies on the perpendicular bisector of AB, $OB = OA = R$ by Proposition 2.4, and as O lies on the perpendicular bisector of AC, $OC = OA = R$ as well.) So this circle passes through the three vertices A, B, C of the triangle, and is called the *circumcircle* of $\triangle ABC$. The radius R of the circumcircle is called the *circumradius* of $\triangle ABC$.

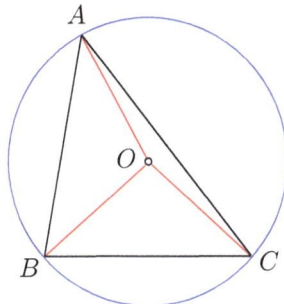

Theorem 2.3. *In any triangle, the perpendicular bisectors of its sides are concurrent.*

Proof. Let the perpendicular bisectors of the sides AB, CA meet at a point O. We want to show that O also lies on the perpendicular bisector of BC.

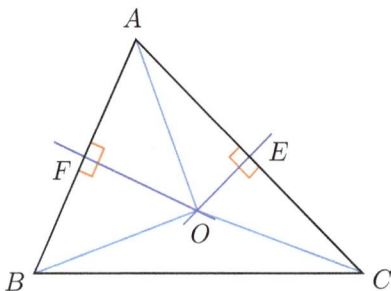

As O lies on the perpendicular bisector of AB, we have $OA = OB$. Similarly, since O also lies on the perpendicular bisector of AC, we also have $OA = OC$. It follows that $OB = OC$. But by the Perpendicular Bisector Theorem, this implies that O lies on the perpendicular bisector of BC. Thus O is a common point for the perpendicular bisectors of the three sides of the triangle.   □

2.3   ASA and AAS Congruency Rules

Theorem 2.4 (ASA Congruency Rule). *If two angles and the included side of one triangle are equal to the corresponding two angles and the included side of the other triangle, then the two triangles are congruent.*

Proof. Name the triangles $\triangle ABC$ and $\triangle A'B'C'$, such that $\angle B = \angle B'$, $\angle C = \angle C'$ and $BC = B'C'$. If $AB = A'B'$, then we are done by the SAS Congruency Rule. So let us suppose that $AB \neq A'B'$. We may assume that $AB < A'B'$. (The proof when $AB > A'B'$ is analogous, and is obtained by simply swapping the primed and unprimed vertices below!) Let A'' be a point on $A'B'$ such that $A''B' = AB$.

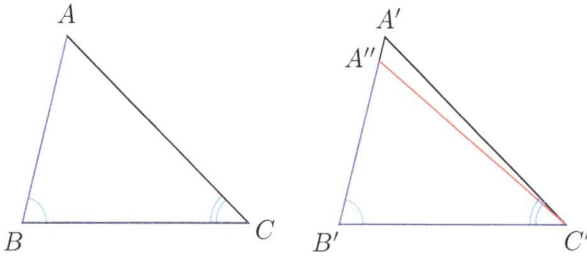

In $\triangle ABC$ and $\triangle A''B'C'$, $AB = A''B'$, $BC = B'C'$ and $\angle B = \angle B'$. Hence by the SAS Congruency Rule, $\triangle ABC \simeq \triangle A''B'C'$. So $\angle C = \angle A''C'B'$. But $\angle C = \angle A'C'B'$ (given). Hence

$$\angle A'C'A'' = \angle A'C'B' - \angle A''C'B' = \angle C - \angle C = 0°,$$

that is, A'' coincides with A', and so $AB = A'B'$. $\qquad\qquad\square$

Corollary 2.2 (AAS Congruency Rule). *If two angles and a nonincluded side of one triangle are equal to the corresponding two angles and the corresponding side of the other triangle, then the two triangles are congruent.*

Proof. The sum of the three angles in any triangle is constant ($180°$), and so the equality of two corresponding angles in the two triangles automatically gives the equality of the remaining third angles in the two triangles. Then by the ASA Congruency Rule, the two triangles are congruent. $\quad\square$

Corollary 2.3. *The lengths of perpendicular segments between a pair of parallel lines is the same.*

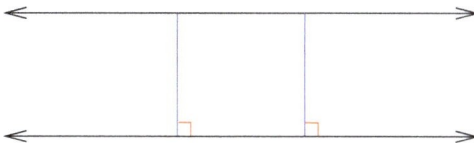

Proof. Let the two lines be ℓ, ℓ'. Suppose that A, B be points on ℓ and A', B' be points on ℓ' such that the interior angles at A, B are $90°$. Then all the interior angles are equal to $90°$. Join $A'B$.

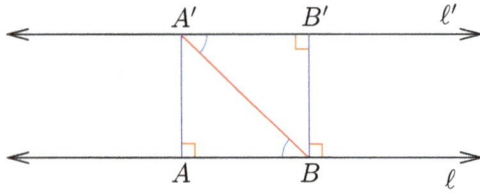

Then in the triangles $\triangle AA'B$ and $\triangle B'BA'$, we have

$$A'B = A'B \quad \text{(common side)},$$
$$\angle A'AB = 90° = \angle BB'A',$$
$$\angle A'BA = \angle BA'B' \quad \text{(alternate interior angles, } \ell' \parallel \ell\text{)}.$$

By the AAS Congruency Rule, $\triangle AA'B \simeq \triangle B'BA'$. So $A'A = B'B$ (CPCT). □

Corollary 2.4. *Any point on the angle bisector of an angle is equidistant from the bounding rays of the angle.*

Proof. Let the angle have the vertex O, and let P be a point on the angle bisector. Drop perpendiculars PX, PY on the two bounding rays of the angle.

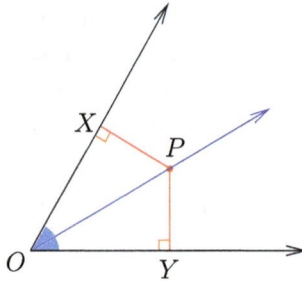

We will show that $\triangle OPX \simeq \triangle OPY$. We have

$$OP = OP \quad \text{(common side)},$$
$$\angle XOP = \angle YOP \quad \text{(given)},$$
$$\angle OXP = \angle OYP = 90° \quad \text{(construction)}.$$

By the AAS Congruency Rule, $\triangle OPX \simeq \triangle OPY$. Hence $PX = PY$. (CPCT). □

In Proposition 2.1, we had learnt that the opposite angles to the equal sides in an isosceles triangle are equal. Now we will prove the converse.

Corollary 2.5. *In a triangle with two equal angles, the sides opposite them are equal, that is, the triangle is isosceles.*

Proof. Let $\triangle ABC$ be a triangle with $\angle B = \angle C$. Consider the angle bisector of the angle $\angle BAC$, meeting the side BC at the point D.

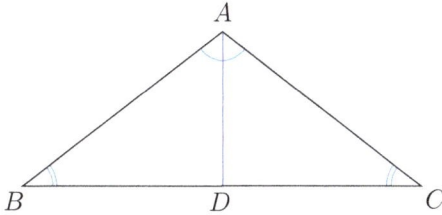

We will show that $\triangle BAD \simeq \triangle CAD$. We have

$$AD = AD \quad \text{(common side)},$$
$$\angle BAD = \angle CAD \quad \text{(construction)},$$
$$\angle ABD = \angle ACD \quad \text{(given)}.$$

By the AAS Congruency Rule, $\triangle BAD \simeq \triangle CAD$. Thus $AB = AC$ (CPCT).

A slicker proof is as follows. In $\triangle ABC$ and $\triangle ACB$, we have that $\angle B = \angle C$. Also the angle $\angle A$ is common to both, and so is the side BC. By the AAS Congruency Rule, $\triangle ABC \simeq \triangle ACB$, and so $AB = AC$ (CPCT). □

Exercise 2.9. (∗) (This exercise should be attempted after the discussion of areas and the Midpoint Theorem.) Through a given point within a given angle, construct a line which forms a triangle of minimum area.

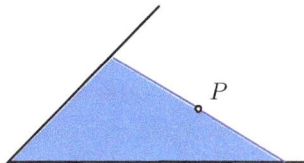

2.4   SSA and RHS Congruency Rules

We have seen that when we have a pair of equal corresponding angles in two triangles with an equal side in both, it doesn't matter whether the side is included between the angles, that is, we have both the ASA and the AAS Congruency Rules. On the other hand, we have the SAS Congruency Rule, and this prompts the question of whether one can also expect an SSA Congruency Rule? We had seen earlier that the answer is no in general. Nevertheless we have the following result.

Theorem 2.5 (SSA Congruency Rule). *Let $\triangle ABC$ and $\triangle A'B'C'$ be triangles such that*

(1) $AB = A'B'$
(2) $AC = A'C'$
(3) $\angle ABC = \angle A'B'C'$.

Then either $\triangle ABC \simeq \triangle A'B'C'$

     or $\angle C + \angle C' = 180°$.

Note that if we have additional information in the pair of triangles, then we can sometimes conclude that the two triangles are congruent. For example, if we happen to know that $\angle C$ and $\angle C'$ are both strictly less than $90°$, then we know that $\angle C + \angle C' \neq 180°$, and so we must then have with the hypothesis of the above theorem that $\triangle ABC \simeq \triangle A'B'C'$.

Proof. Let us consider the following two cases.

1° $\triangle ABC \simeq \triangle A'B'C'$. Then we are done.

2° $\triangle ABC \not\simeq \triangle A'B'C'$. We claim that then we must have $BC \neq B'C'$. For otherwise if $BC = B'C'$, then all the corresponding sides in the triangles $\triangle ABC$ and $\triangle A'B'C'$ would be equal, so that by the SSS Congruency Rule the two triangles would be congruent, which is false in the case we are considering! So $BC \neq B'C'$. We may assume without loss of generality that $BC > B'C'$. (Otherwise we can carry out the following argument by making the swaps $A \leftrightarrow A'$, $B \leftrightarrow B'$, $C \leftrightarrow C'$.) Mark the point C'' on BC such that $BC'' = B'C'$.

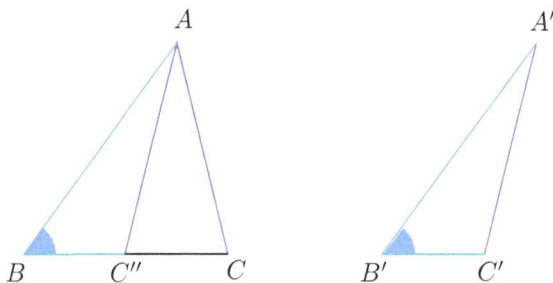

By the SAS Congruency Rule, $\triangle ABC'' \simeq \triangle A'B'C'$. It follows that $\angle AC''B = \angle A'C'B'$ (CPCT). Also $AC'' = A'C' (= AC)$. Thus $\triangle AC''C$ is isosceles, and $\angle AC''C = \angle ACC''$. So

$$\angle C + \angle C' = \angle AC''C + \angle AC''B = 180°.$$

This completes the proof. □

Corollary 2.6 (RHS Congruency Rule). *A pair of right angled triangles having the same lengths of their hypotenuses and same lengths of one other side are congruent.*

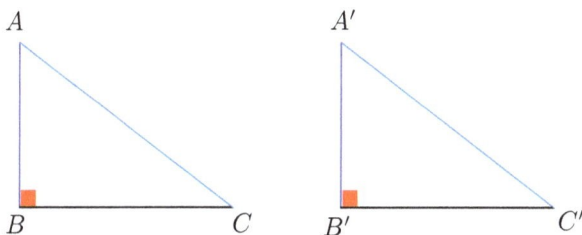

Proof. This is a consequence of the SSA Congruency Rule. Suppose the two triangles are $\triangle ABC$ and $\triangle A'B'C'$, with the right angles at B, B'. Then $\angle C$ and $\angle C'$ are both strictly less than $90°$, and so $\angle C + \angle C' < 180°$. By Theorem 2.5, it follows that $\triangle ABC \simeq \triangle A'B'C'$. □

In Corollary 2.4 we learnt that points on the angle bisector are equidistant from the rays defining the angle. Now we will learn the converse.

Corollary 2.7. *If a point in the interior of an angle is equidistant from the two rays defining the angle, then it must lie on the angle bisector of the angle.*

Proof. Let the angle have vertex O, and suppose that the point P in the interior of the angle is equidistant from the two rays bounding the angle. Drop perpendiculars from the point P to the two rays bounding the angle, and suppose they meet the two rays at the points X and Y. Then we know that $PX = PY$. Join O to P. We want to show that OP bisects angle XOY, that is, $\angle XOP = \angle YOP$. We'll do this by showing $\triangle XOP \simeq \triangle YOP$.

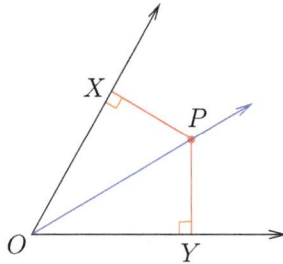

By the RHS Congruency Rule, $\triangle XOP \simeq \triangle YOP$, since OP is a common side, $PX = PY$ (given), and $\angle PXO = \angle PYO = 90°$ (construction). Consequently $\angle XOP = \angle YOP$ (CPCT). This completes the proof.   □

Exercise 2.10. (All triangles are equilateral!) The aim of this exercise is to show that all triangles are equilateral! Let $\triangle ABC$ be the triangle. Construct the perpendicular bisector of side BC, the angle bisector of $\angle BAC$, and let O be their point of intersection. Drop perpendiculars from the point O to the sides AB and AC, meeting them at the points P, N, respectively. Join O to B and also to C.

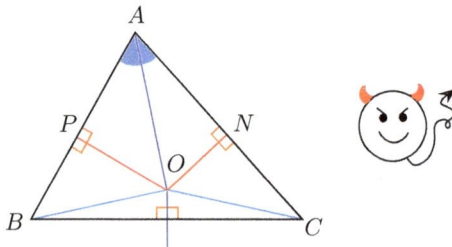

In $\triangle APO$ and $\triangle ANO$, the side AO is common, $\angle APO = \angle ANO = 90°$, and $\angle PAO = \angle NAO$. Hence by the AAS Congruency Rule, $\triangle APO \simeq \triangle ANO$. Thus $AP = AN$ (CPCT).

   In $\triangle POB$ and $\triangle NOC$, $OP = ON$ (since we know that $\triangle APO \simeq \triangle ANO$), $\angle BPO = \angle CNO = 90°$, and $OB = OC$ (as O lies on the perpendicular bisector of BC). Hence by the RHS Congruency Rule, $\triangle POB \simeq \triangle NOC$. It follows that $PB = NC$ (CPCT).

From the above we have $AB = AP + PB = AN + NC = AC$. So we have got that in any triangle $\triangle ABC$, $AB = AC$. Similarly, $BC = AC$. Consequently, $\triangle ABC$ is equilateral.

Clearly there is a fallacy in the above argument. Where does it lie?

2.5   Angle bisectors in a triangle are concurrent

We will now show that the angle bisectors in a triangle are concurrent. The point of concurrency is called the *incenter* of the triangle. As the incenter lies on the angle bisectors of each of the three angles, it is equidistant from their bounding rays, and so it is a point equidistant from the sides of a triangle. So if we drop perpendiculars from the incenter I to the three sides of the triangle, meeting the sides at the points D, E, F, then $ID = IE = IF =: r$. Hence if we draw a circle with radius equal r, then we obtain a circle which is called the *incircle* of the triangle, touching the sides at D, E, F. We will see later on in the chapter on circles that the three sides are "tangential" to the incircle, that is, each of the three sides meet the circle at exactly one point (D, E, F respectively). The radius r of the incircle is called the *inradius* of the triangle.

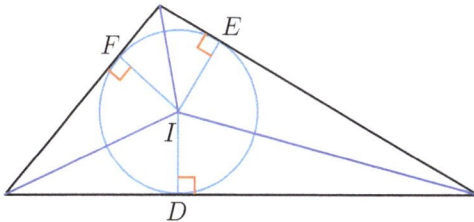

Theorem 2.6. *In any triangle, the angle bisectors of its three angles are concurrent.*

Proof. In a triangle $\triangle ABC$, let the angle bisectors of the angles B and C meet at a point I. Join A to I. We wish to show that AI bisects angle $\angle A$. Drop perpendiculars from I to the three sides, and let them meet BC, CA, AB in D, E, F, respectively.

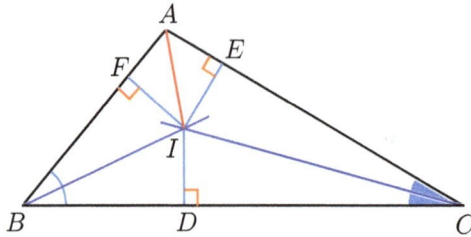

Since the point I lies on the angle bisector of $\angle B$, it follows from Corollary 2.4 that $IF = ID$. Similarly, since I lies on the angle bisector of $\angle C$, also $ID = IE$. Hence $IF = (ID =)IE$. But now the point I in the interior of $\angle A$ is equidistant from its bounding rays, and so by Corollary 2.7, we conclude that I lies on the angle bisector of $\angle A$. This means that the three angle bisectors meet at the point I.   □

Exercise 2.11. (∗) Suppose that f is a real valued function defined in the plane such that f has the property that for any triangle $\triangle ABC$ with incenter I,

$$f(I) = \frac{f(A) + f(B) + f(C)}{3}.$$

Show that f must be constant.

Chapter 3

Quadrilaterals

Definition 3.1 (Quadrilateral). A *quadrilateral* is a figure with 4 sides and 4 vertices.

Loosely speaking, a *side* is a line segment subset of the planar figure which is maximally contained in the figure; in other words, if we add any more points to the subset by extending the line segment, then we fall outside the figure. A *vertex* is a point where two distinct sides of the figure meet.

Trapezium	A pair of opposite sides are parallel	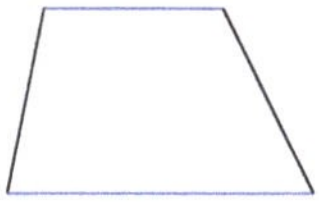
Parallelogram	Opposite sides are parallel	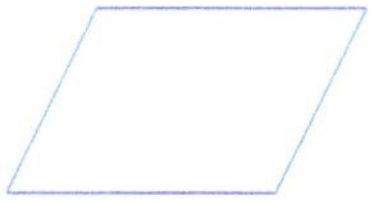
Rectangle	Parallelogram with an angle = $90°$	
Square	Rectangle with equal sides	
Rhombus	Quadrilateral with equal sides	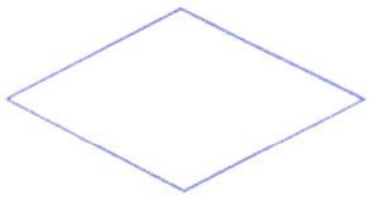

Exercise 3.1. Make sense of the statement "Prime numbers are nonrectangular numbers".

3.1   Characterizations of a parallelogram

Proposition 3.1 (Properties/Characterizations of a Parallelogram).
The following statements are equivalent:

(1) *ABCD is a parallelogram.*
(2) *$AB = CD$ and $AD = BC$. (Opposite sides are equal.)*
(3) *$\angle A = \angle C$ and $\angle D = \angle B$. (Opposite angles are equal.)*
(4) *$AB = CD$ and $AB \parallel CD$. (A pair of opposite sides are equal and parallel.)*

Proof. We'll show that $(1) \Rightarrow (2) \Rightarrow (3) \Rightarrow (4) \Rightarrow (1)$.

$(\mathbf{1}) \Rightarrow (\mathbf{2})$. Join A to C. Consider the two triangles ADC and CBA. We will show that they are congruent. As $AD \parallel BC$, the alternate angles to the transversal AC are equal, giving $\angle DAC = \angle ACB$. Also, $CD \parallel AB$, and so $\angle DCA = \angle CAB$, as they are alternate to the transversal AC. Finally the side AC is common to $\triangle ADC$ and $\triangle CBA$. Thus by the ASA Congruency Rule, $\triangle ADC \simeq \triangle CBA$. Hence $AD = BC$ and $AB = CD$ (CPCT). Thus (2) holds.

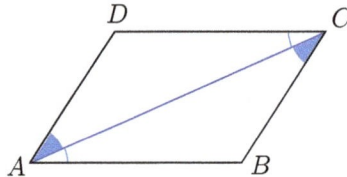

$(\mathbf{2}) \Rightarrow (\mathbf{3})$. Join A to C. Consider the two triangles ADC and CBA. We will show that they are congruent. The side AC is common to the two triangles and it is given that $AD = BC$ and $AB = CD$. Thus by the SSS Congruency Rule, $\triangle ADC \simeq \triangle CBA$. Hence $\angle B = \angle D$ (CPCT). Also,

$$\angle DAC = \angle ACB \quad \text{(CPCT), and}$$
$$\angle CAB = \angle DCA \quad \text{(CPCT),}$$

and adding these, we obtain $\angle A = \angle C$. Thus (3) holds.

(3) $\Rightarrow$ **(4)**. As $\angle A + \angle B + \angle C + \angle D = 360°$, we obtain using $\angle A = \angle C$ and $\angle B = \angle D$ that $\angle A + \angle D = 180°$. As these are alternate interior angles for the transversal AD, we conclude that $AB \parallel CD$. Now join A to C.

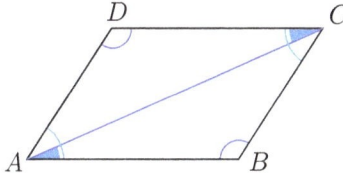

We have $\angle B = \angle D$ (given), and $\angle DCA = \angle BAC$ (alternate angles). More-over, the side AC is common to the two triangles $\triangle DCA$ and $\triangle BAC$. By the AAS Congruency Rule, $\triangle DCA \simeq \triangle BAC$. Hence $AB = CD$ (CPCT). Thus (4) holds.

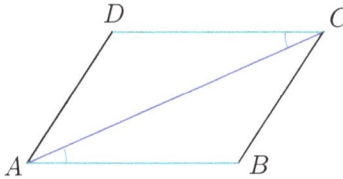

(4) $\Rightarrow$ **(1)** Join A to C. Consider the two triangles $\triangle ADC$ and $\triangle CBA$. We will show that they are congruent. The side AC is common to the two triangles and it is given that $AB = CD$. Moreover, since $AB \parallel CD$, the alternate angles to the transversal AC are equal, giving $\angle DCA = \angle CAB$. Thus by the SAS Congruency Rule, $\triangle ADC \simeq \triangle CBA$. Hence we have $\angle DAC = \angle BCA$ (CPCT). It follows that $AD \parallel BC$. Consequently $ABCD$ is a parallelogram. $\square$

Exercise 3.2. Show that a quadrilateral is a parallelogram if and only if its diagonals bisect each other.

Exercise 3.3. Show that each median[1] of a triangle has a length that is at most the average of the lengths of the adjacent sides.

---

[1]A line joining the vertex of a triangle to the midpoint of the opposite side is called a *median*.

Exercise 3.4. (Three equal circles).(∗) If three circles having the same radius pass through a point, then the other points of intersections of these circles lie on a circle of the same radius.

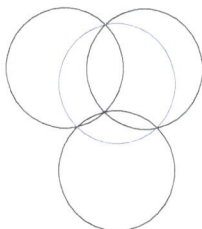

Exercise 3.5. Let f be a real-valued function on the plane such that for every square $ABCD$ in the plane, $f(A) + f(B) + f(C) + f(D) = 0$. Show that $f \equiv 0$, that is, f is constant, taking the value 0 everywhere in the plane.

Altitudes of a triangle are concurrent

In a triangle, a line segment from a vertex of the triangle to a point on the line containing the opposite side which is moreover perpendicular to it is called an *altitude*.

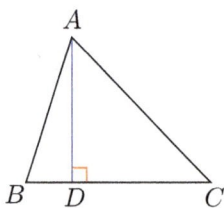

AD is an altitude.

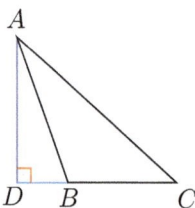

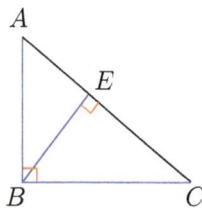

The altitudes are
AB, BC, BE.

As we can always "drop a perpendicular" from a point outside a line to the line (Construction 2.6), we see that there are three altitudes in a triangle, one from each vertex. We will show below that, remarkably, the three altitudes of a triangle are concurrent. The point of intersection of the three altitudes of a triangle is called the *orthocenter* of the triangle.

Theorem 3.1. *In any triangle, the altitudes are concurrent.*

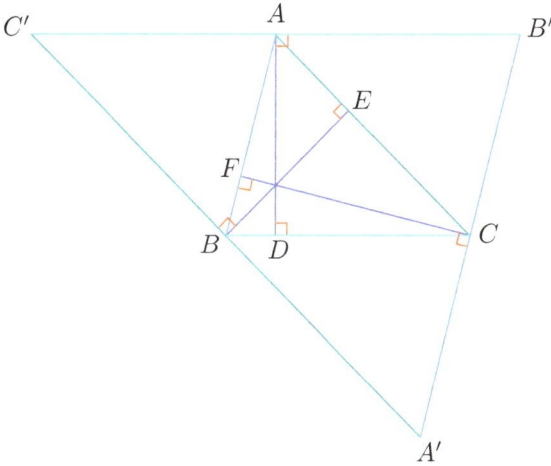

Proof. Let AD, BE, CF be the altitudes of the triangle $\triangle ABC$. Draw a line ℓ_A through A parallel to BC, and similarly lines ℓ_B, ℓ_C through B, C, respectively, parallel to AC, AB respectively. Let ℓ_A, ℓ_B meet at C', ℓ_B, ℓ_C meet at A', and finally, ℓ_A, ℓ_C meet at B'. Then $C'ACB$, $B'CBA$ are both parallelograms, and so in each, pairs of opposite sides are equal, giving in particular $C'A = BC = B'A$. So A is the midpoint of $B'C'$. Hence AD is the perpendicular bisector of the side $B'C'$ in $\triangle A'B'C'$. Similarly, we see that BE, CF are perpendicular bisectors of the sides $C'A'$ and $A'B'$, respectively, of $\triangle A'B'C'$. But by Theorem 2.3, we know that the perpendicular bisectors of any triangle are concurrent. Thus the perpendicular bisectors AD, BE, CF of $\triangle A'B'C'$ are concurrent. But these happen to be the altitudes of $\triangle ABC$. Consequently the altitudes AD, BE, CF of $\triangle ABC$ are concurrent. $\square$

Midpoint theorem

Theorem 3.2 (Midpoint Theorem). *In any triangle, the line segment joining the midpoints of two sides is parallel to the third side, and is half its length.*

Proof. Let D, E be midpoints of AB, AC, respectively. Extend DE to a point F such that $DE = EF$. Join C to F.

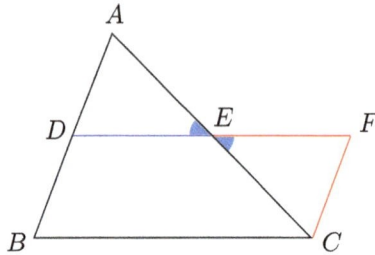

Consider $\triangle ADE$ and $\triangle CFE$. We will show that they are congruent. We have

$$\angle AED = \angle CEF \quad \text{(opposite angles)}$$
$$AE = EC \quad \text{(E is the midpoint of AC)}$$
$$DE = EF \quad \text{(construction of F).}$$

By the SAS Congruency Rule $\triangle ADE \simeq \triangle CFE$. Thus $BD = AD = CF$ and $\angle DAE = \angle FCE$ (CPCT). But $\angle DAE$ and $\angle FCE$ are alternate angles to the transversal AC intersecting the two lines AB and CF, and these angles being equal, we conclude that the lines AB and CF are parallel. Thus the quadrilateral $BCFD$ is a parallelogram since a pair of opposite sides are equal and parallel. Consequently, $DE \parallel BC$. Moreover, we have $DE = DF/2 = BC/2$. $\hspace{2cm}$ $\square$

One can also show the following converse result.

Theorem 3.3 (Midpoint theorem 2). *In any triangle, a line drawn through the midpoint of one side and parallel to another side meets the third side in its midpoint.*

Proof. Let D be the midpoint of AB, and draw the line DE parallel to BC, meeting AC in E. We'd like to show that E is the midpoint of AC.

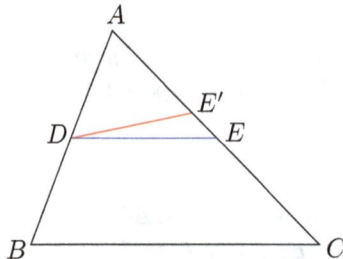

Suppose on the contrary, that E' is the midpoint of AC and E' is distinct from E. Join D to E'. Then by the midpoint theorem $DE' \parallel BC$. But also $DE \parallel BC$. So we have a point (D) outside the line ℓ containing BC, through which we are able to draw two distinct parallel lines (namely the line containing DE and the line containing DE'). This contradicts the Parallel Postulate. □

Exercise 3.6. Show that in any quadrilateral, the quadrilateral obtained by joining the midpoints of the sides is a parallelogram.

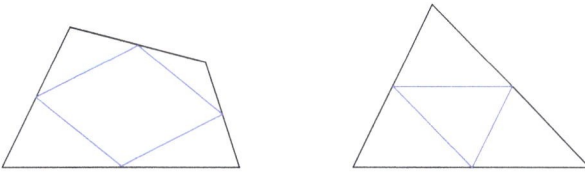

Exercise 3.7. Show that in any triangle, the four smaller triangles obtained by joining the midpoints of the sides are all congruent to each other.

Exercise 3.8. Show that in any parallelogram $ABCD$, if Q, S are the midpoints of the sides BC and DA, then the lines DQ and BS trisect the diagonal AC.

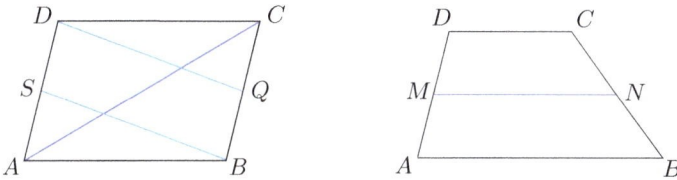

Exercise 3.9. In a trapezium $ABCD$, where $AB \parallel CD$, let M, N be the midpoints of the sides AD, BC, respectively. Prove that $MN = (AB + CD)/2$.

Exercise 3.10. Prove that the line segment joining the midpoints of the diagonals of a trapezium is parallel to the parallel sides and equal to half their difference.

Exercise 3.11. Show, using the following picture, that every triangle can be dissected into four isosceles triangles.

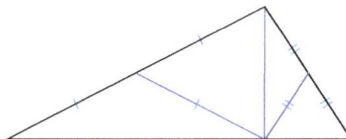

The medians of a triangle are concurrent

A line joining the vertex of a triangle to the midpoint of the opposite side is called a *median*.

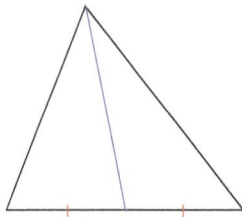

Clearly there are three medians in a triangle, corresponding to each of the vertices. It turns out that these are concurrent, as we will show below. Their meeting point is called the *centroid* of the triangle.

Theorem 3.4. *In any triangle, the three medians are concurrent. Moreover, their common intersection point (the centroid) divides each median in the ratio* $2:1$.

Proof. Let BE, CF be two medians of the triangle $\triangle ABC$ meeting at the point G. Join AG, and extend it to meet BC in D. We want to show that D is the midpoint of the side BC (so that AD is the median, and then all three medians pass through G, which will be the centroid of the triangle).

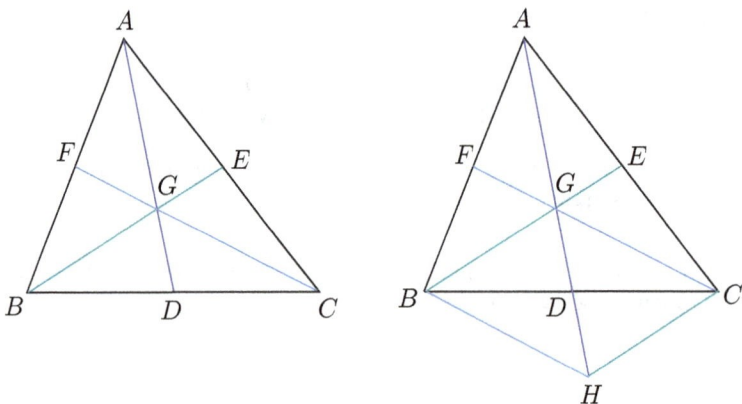

Extend AG to a point H such that $GH = AG$. Consider $\triangle ABH$. As F, G

are midpoints of the sides AB, AH, respectively, it follows by the Midpoint Theorem that $FG \parallel BH$. So $GC \parallel BH$.

Next consider $\triangle ACH$. As G, E are the midpoints of the sides AH, AC, respectively, again by the Midpoint Theorem, $GE \parallel CH$, that is, $BG \parallel HC$.

Thus in the quadrilateral $BHCG$, opposite sides are parallel, and so it is a parallelogram. But in a parallelogram, the diagonals bisect each other (Exercise 3.2). Hence D is the midpoint of BC. This completes the proof of the first claim.

To show that the centroid G divides each median in the ratio $2 : 1$, we proceed as follows.

(1) By construction, $AG = GH$, and as D is the midpoint of the diagonal of the parallelogram $BCGH$, also $GH = 2GD$. Hence $AG : GD = 2 : 1$.
(2) By the Midpoint Theorem in $\triangle ABH$, $GF = BH/2$. But as $BGCH$ is a parallelogram, its opposite sides are equal, and so $BH = CG$. Thus we have $CG : GF = 2 : 1$.
(3) Similarly, by the Midpoint Theorem in $\triangle AHC$, $GE = HC/2$. But as $BGCH$ is a parallelogram, its opposite sides are equal, and so we have $HC = BG$. Hence $BG : GE = 2 : 1$.

$\square$

Exercise 3.12. Given a triangle $\triangle ABC$, a new triangle $\triangle A_1 B_1 C_1$ is formed by joining the midpoints of the sides. This process is repeated indefinitely, resulting in a sequence of shrinking triangles, converging to a point P. Describe P in terms of $\triangle ABC$.

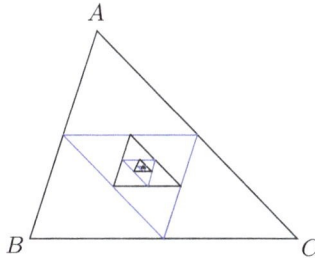

3.2   Areas

Just like we can measure lengths of line segments by comparison with a chosen unit length line segment, we can compare the areas of regions in the plane by comparison with a region of "unit area". The simplest region one can think of for easy comparison is a *unit square*, which is a square with unit side lengths. We abbreviate this measure of "unit area" as a *square unit*. The first picture from the left below shows a unit area square, which has the area 1 square unit. The middle picture shows a rectangle with side lengths 2 and 3, and by tiling with squares of unit length, we see that its area must be $(2) \cdot (3) = 6$ square units.

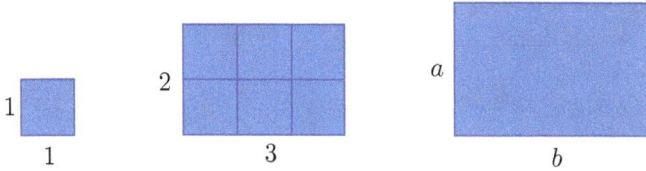

Now suppose that the rectangle had rational lengths $a := \frac{p}{q}$ and $b := \frac{m}{n}$ units, where $p, q, m, n \in \mathbb{N}$. We will justify that its area is $a \cdot b = \frac{p}{q} \cdot \frac{m}{n}$. To see this, we first divide our unit square into $(qn)^2$ congruent little squares (achieved by dividing each unit length side into qn equal parts). Now we will tile our big rectangle with side lengths a, b with these little squares (each having a side length of $\frac{1}{qn}$, and each having an area of $\frac{1}{(qn)^2}$. Clearly pn such little tiles can be placed along the side length a, and mq can be placed along b. Consequently, the total number of little tiles needed to cover the big rectangle is $(mq)(pn)$, and so the area of the big rectangle is

$$(mq) \cdot (pn) \cdot \frac{1}{(qn)^2} = \frac{p}{q} \cdot \frac{m}{n} = a \cdot b.$$

Based on this, if a, b are any two real numbers, then we define the area of the rectangle with side lengths a, b as $a \cdot b$. See the rightmost figure above.

Lemma 3.1. *The area of a right angled triangle with nonhypotenuse side lengths a, b is $(ab)/2$.*

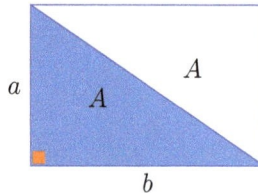

Proof. Just consider a rectangle with side lengths a, b. By drawing a diagonal, we see that we get two congruent triangles with nonhypotenuse side lengths a, b, and so the area of each triangle is half the area $a \cdot b$ of the rectangle. □

Using this result, we can now tackle the case of general triangles.

Proposition 3.2. *The area of a triangle is half the product of the length of any side of the triangle and the length of the altitude to the chosen side.*

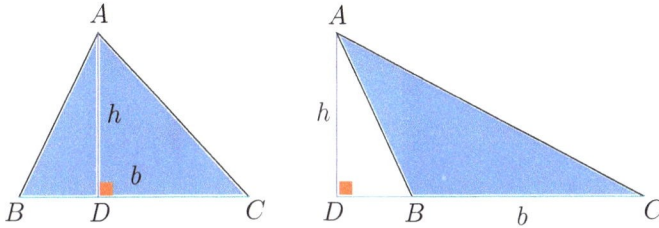

The chosen side is referred to as a *base* and the length of the altitude to the base is referred to as the *height*. We will denote the area of a triangle $\triangle ABC$ simply by $\triangle ABC$, analogous to our previous notation AB for the length of a line segment AB and $\angle ABC$ for the measure of angle $\angle ABC$.

Proof. Let $\triangle ABC$ be a triangle, and BC be the chosen side. Suppose first that the altitude from the vertex A meets the side BC at a point D *between* B and C. Then the area of $\triangle ABC$ is the sum of the areas of the two *right angled* triangles $\triangle ADB$ and $\triangle ADC$, and so

$$\triangle ABC = \triangle ADB + \triangle ADC = \frac{BD \cdot h}{2} + \frac{DC \cdot h}{2}$$
$$= \frac{BD + DC}{2} \cdot h$$
$$= \frac{BC \cdot AD}{2}.$$

On the other hand, if the altitude to the base BC falls outside BC, then rather than the sum, we would need to take the difference of the areas of $\triangle ADB$ and $\triangle ADC$. Referring to the picture on the right above, we have in this case

$$\triangle ABC = \triangle ADC - \triangle ADB = \frac{DC \cdot h}{2} - \frac{DB \cdot h}{2}$$
$$= \frac{DC - DB}{2} \cdot h$$
$$= \frac{BC \cdot AD}{2}.$$

This completes the proof. $\square$

Thus given a fixed base BC and a fixed line ℓ parallel to BC, the area of each triangle $\triangle ABC$ with vertex A lying on ℓ is the same, namely $(BC \cdot h)/2$, where h is the distance between ℓ and BC. Indeed, all these triangles have the same length h of the altitude drawn from A to the base BC.

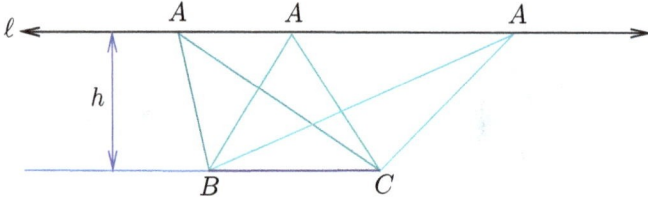

Theorem 3.5 (Angle Bisector Theorem). *In a triangle, the bisector of an angle divides the opposite side in the ratio of the sides bounding the angle.*

Proof. Let the bisector of $\angle A$ in the triangle $\triangle ABC$ meet the opposite side BC at a point D. We want to show that

$$\frac{BD}{DC} = \frac{AB}{AC}.$$

Drop perpendiculars from D to the sides AB, AC, meeting them (or possibly their linear extensions) at X, Y, respectively. Also, drop a perpendicular from A to the side BC, meeting it at Z.

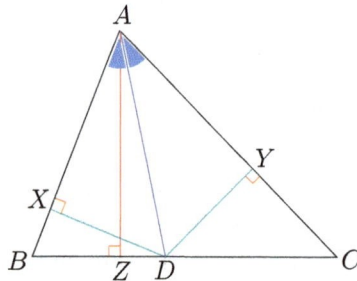

Then DX, AZ are altitudes in $\triangle ABD$, and DY, AZ are altitudes in $\triangle ADC$. We note that $DX = DY$, since D, being a point lying on the angle bisector of $\angle A$, is equidistant from the rays bounding $\angle A$. Thus we have that the ratio of the areas of these two triangles is

$$\frac{\triangle ABD}{\triangle ADC} = \frac{\frac{1}{2}AB \cdot DX}{\frac{1}{2}AC \cdot DY} = \frac{\frac{1}{2}BD \cdot AZ}{\frac{1}{2}DC \cdot AZ}.$$

$\square$

Exercise 3.13. Prove the converse of the Angle Bisector Theorem. That is, show that if D is on the side BC of $\triangle ABC$ such that $AB : AC = BD : DC$, then AD bisects $\angle A$.

Exercise 3.14. Show that the area of a parallelogram is equal to the length of one of the sides and the distance between this side and the side parallel to it.

Exercise 3.15. In a triangle $\triangle ABC$, altitudes AD, BE, CF are drawn. If $AB : BC : CA = 4 : 5 : 6$, then show that $CF : AD : BE = \dfrac{1}{4} : \dfrac{1}{5} : \dfrac{1}{6}$.

Exercise 3.16. Let r be the inradius r of a triangle $\triangle ABC$ with area Δ and side lengths a, b, c. Show that

$$r = \frac{2\Delta}{a + b + c}.$$

Exercise 3.17. Let h_A, h_B, h_C be the altitudes lengths in a triangle, and r be its inradius. By finding the area of a triangle in two different ways, prove that

$$\frac{1}{h_A} + \frac{1}{h_B} + \frac{1}{h_C} = \frac{1}{r}.$$

Exercise 3.18. (Hocus pocus!) Consider the 8×8 square below which has been cut into four pieces and reassembled to form a 5×13 rectangle. So the original area of 64 square units has been changed to 65 square units!

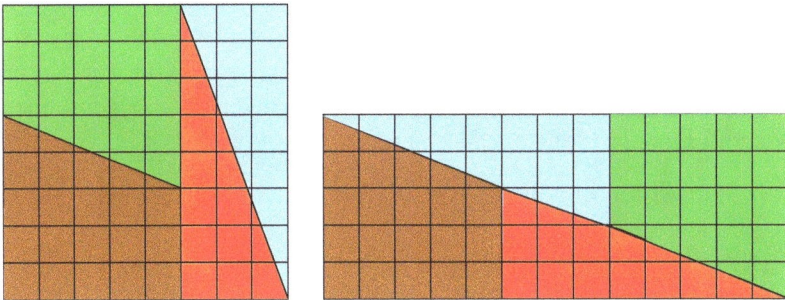

Was the wool pulled over your eyes or did we manage to create an extra square unit area by just a rearrangement?

Exercise 3.19. Let the vertices A, B, C of a right angled triangle of unit area be reflected in their respective opposite sides to obtain new points A', B', C'. What is the area of $\triangle A'B'C'$?

Exercise 3.20. By considering the staircase below, where the length of the room is $n+1$ units and the height is n units, and each step has uniform width and height, show that for all natural numbers n,

$$1 + 2 + \cdots + n = \frac{n(n+1)}{2}.$$

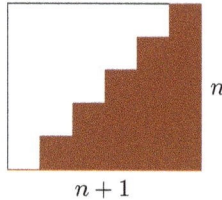

n

$n+1$

Exercise 3.21. (Viviani's[2] Theorem). Show that in any equilateral triangle, the sum of the distances of a point inside it to the three sides equals the length of the altitude.

Exercise 3.22. On the sides AB, BC, CA of a triangle $\triangle ABC$, points C', A', B', respectively, are such that $AC' : C'B = BA' : A'C = CB' : B'A = 2 : 1$. Determine the ratio of the areas of the two triangles $\triangle A'B'C'$ and $\triangle ABC$. Repeat the exercise more generally when $AC' : C'B = BA' : A'C = CB' : B'A = x > 0$. Do you get the expected answer when $x = 0$ and when $x = 1$?

Kepler's Second Law

Based on the astronomical observations of the planets in our solar system made by the Danish astronomer Tycho Brahe (1546-1601), the German mathematician and astronomer Johannes Kepler (1571-1630) formulated (Kepler's) Laws of Planetary Motion:

- (K1) The orbit of a planet is an ellipse[3] with the Sun at one of the two foci[4].
- (K2) A line segment joining a planet and the Sun sweeps out equal areas during equal intervals of time.
- (K3) The square of the orbital period of a planet is proportional to the cube of the semi-major axis of its (elliptic) orbit.

Although Kepler's laws were correct, he was unable to explain *why* the planets behaved the way they did. This was accomplished by the English

---

[2] Due to the 17th century Italian mathematician Vincenzo Viviani.
[3] For the definition of an ellipse, see page 112.
[4] See page 112.

theoretical physicist Issac Newton (1642-1726) as part of his *Principia*, published in 1687, in which he *deduced* Kepler's laws based on his theory of motion and of gravity.

The aim of this paragraph is to derive the second law of Kepler based on Newton's dynamics. Newton's first two laws of motion are:

(N1) If no forces are acting on a body, then it will either stay at rest or continue to travel in a straight line at a constant speed.

(N2) The change in the velocity of a body is proportional to and in the direction of the force acting on the body.

Using these two laws, Newton showed why Kepler's observation, that the planets sweep out equal areas in equal times, is true.

Suppose that a planet travels from A to B in a certain amount of time δt.

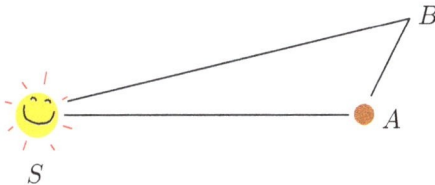

If no force were acting in the planet, then it would continue to a point C' after another unit of time δt. However, the gravitational pull towards the sun makes the planet move to the point C instead, thanks to the impulse BV it receives, directed along BS. So C is the "vector sum" of AB with BV, that is, $BVCC'$ is a parallelogram. See the picture below.

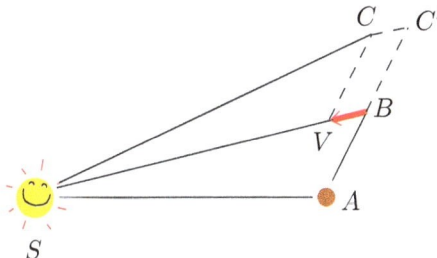

Note that in order to show Kepler's Second Law, we must show that we have equality of the areas $\triangle SAB = \triangle SBC$. Let us first prove that the area $\triangle SAB$ equals the area $\triangle SBC'$. This is clear from the picture below, since the two little right triangles are congruent by the AAS Congruency Rule ($AB = BC'$, and two pairs of angles — the right angles, and the opposite angles — are equal), giving

$$2\Delta SAB = SB \cdot (\text{height}) = 2\Delta SBC'.$$

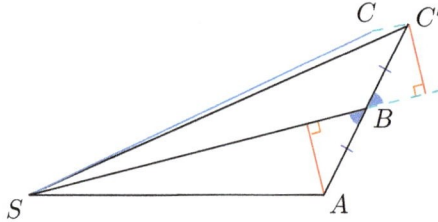

Finally, the proof is completed by showing that $\Delta SBC = \Delta SBC'$. This is because the two triangles ΔSBC and $\Delta SBC'$ have the same base (SB), and the vertices C, C' lie on a line parallel to the base SB.

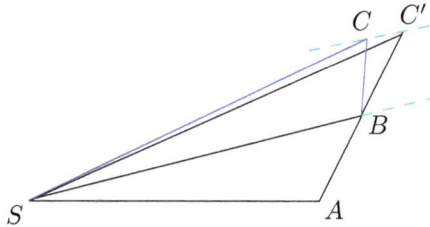

Letting δt shrink to 0, and summing, we get the result for the smooth trajectory of the planet, rather than the discretized version (obtained by breaking the orbit into small straight line segments, giving triangular areas as considered above).

Note that we haven't used the fact that gravitational pull of the Sun is inversely proportional to the square of the radius! All that the proof needs is for the force to be a central one.

Ceva's Theorem

We will now learn about a useful result relating to concurrency in triangles, and we will see that it unifies many known results on concurrency such that that of medians, angle bisectors, altitudes and so on. The result is often attributed to the 17th century Mathematician Giovanni Ceva.

Theorem 3.6 (Ceva's Theorem). *If in triangle* $\triangle ABC$, *points* D, E, F *on the sides* BC, CA, AB, *respectively, are such that* AD, BE, CF *are concurrent, then*

$$\frac{BD}{DC} \cdot \frac{CE}{EA} \cdot \frac{AF}{FB} = 1.$$

Proof. Consider the left picture in the following figure, where P lies in the interior of $\triangle ABC$. The triangles $\triangle ABD, \triangle ADC$ share the altitude dropped from A to BC. Similarly $\triangle BPD, \triangle PDC$ share the altitude dropped from P to BC.

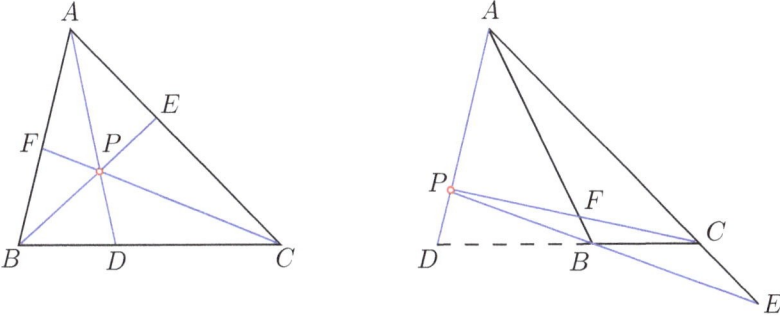

Thus $\dfrac{BD}{DC} = \dfrac{\triangle ABD}{\triangle ADC} = \dfrac{\triangle PBD}{\triangle PDC}.$

(Before we continue, we make the following elementary observation which will be used repeatedly in our proof: Let a, b, α, β be real numbers with b and β nonzero, and such that

$$\frac{\alpha}{\beta} = \frac{a}{b} =: k.$$

If $b + \beta \neq 0$, then also $\dfrac{a+\alpha}{b+\beta} = \dfrac{kb+k\beta}{b+\beta} = \dfrac{k(b+\beta)}{b+\beta} = k.$

Also, as $\dfrac{\alpha}{\beta} = \dfrac{-\alpha}{-\beta}$, the above also gives $\dfrac{a-\alpha}{b-\beta} = k$ if $b - \beta \neq 0$.)

Hence $\dfrac{\triangle APB}{\triangle APC} = \dfrac{\triangle ABD - \triangle PBD}{\triangle ADC - \triangle PDC} = \dfrac{BD}{DC}.$

Similarly $\dfrac{CE}{EA} = \dfrac{\triangle BPC}{\triangle BPA}$ and $\dfrac{AF}{FB} = \dfrac{\triangle CPA}{\triangle CPB}.$

Consequently $\dfrac{BD}{DC} \cdot \dfrac{CE}{EA} \cdot \dfrac{AF}{FB} = \dfrac{\triangle APB}{\triangle APC} \cdot \dfrac{\triangle PBC}{\triangle APB} \cdot \dfrac{\triangle APC}{\triangle PBC} = 1.$

Now let us consider the case when P lies in the exterior of $\triangle ABC$. We refer to the picture on the right in the previous figure. Then

$$\frac{BD}{CD} = \frac{\triangle ADB}{\triangle ACD} = \frac{\triangle PBD}{\triangle PCD} = \frac{\triangle ABD - \triangle PBD}{\triangle ADC - \triangle PDC} = \frac{\triangle APB}{\triangle APC},$$

$$\frac{CE}{EA} = \frac{\triangle PCE}{\triangle APE} = \frac{\triangle BCE}{\triangle ABE} = \frac{-\triangle BCE + \triangle PCE}{-\triangle ABE + \triangle APE} = \frac{\triangle PBC}{\triangle APB},$$

$$\frac{AF}{FB} = \frac{\triangle APF}{\triangle PFB} = \frac{\triangle AFC}{\triangle FCB} = \frac{\triangle APF + \triangle AFC}{\triangle PFB + \triangle FCB} = \frac{\triangle APC}{\triangle PBC}.$$

Hence $\dfrac{BD}{DC} \cdot \dfrac{CE}{EA} \cdot \dfrac{AF}{FB} = \dfrac{\triangle APB}{\triangle APC} \cdot \dfrac{\triangle PBC}{\triangle APB} \cdot \dfrac{\triangle APC}{\triangle PBC} = 1.$ $\square$

Theorem 3.7 (Converse of Ceva's Theorem). *If in triangle $\triangle ABC$, the points D, E, F on BC, CA, AB, respectively satisfy*

$$\frac{BD}{DC} \cdot \frac{CE}{EA} \cdot \frac{AF}{FB} = 1,$$

then AD, BE, CF are concurrent.

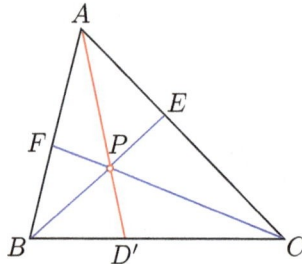

Proof. Let P be the point of intersection of BE and CF. Join A to P and extend it to meet BC in D'. By Ceva's Theorem,

$$\frac{BD'}{D'C} \cdot \frac{CE}{EA} \cdot \frac{AF}{FB} = 1. \tag{3.1}$$

But it is given that

$$\frac{BD}{DC} \cdot \frac{CE}{EA} \cdot \frac{AF}{FB} = 1. \tag{3.2}$$

From (3.1) and (3.2), it follows that $\dfrac{BD}{DC} = \dfrac{BD'}{D'C}$.

Adding 1 to both sides gives $\dfrac{BC}{DC} = \dfrac{BD + DC}{DC} = \dfrac{BD' + D'C}{D'C} = \dfrac{BC}{D'C}$.

Hence $DC = D'C$, and so D coincides with D'. Consequently, AD, BE, CF are concurrent, with the common intersection point P. $\square$

We can now recover the result that the medians in a triangle are concurrent.

Corollary 3.1. *The medians of a triangle are concurrent.*

Proof. As

$$\frac{BD}{DC} = \frac{CE}{EA} = \frac{AF}{FB} = 1,$$

the claim follows by the converse of Ceva's Theorem. □

The concurrency of the angle bisectors of a triangle can also be derived as a consequence of Ceva's Theorem.

Corollary 3.2. *The angle bisectors of a triangle are concurrent.*

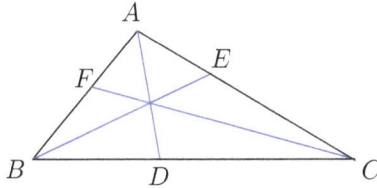

Proof. By the Angle Bisector Theorem,

$$\frac{BD}{DC} = \frac{AB}{AC}, \quad \frac{CE}{EA} = \frac{BC}{AB}, \quad \frac{AF}{FB} = \frac{AC}{BC}.$$

Hence

$$\frac{BD}{DC} \cdot \frac{CE}{EA} \cdot \frac{AF}{FB} = \frac{AB}{AC} \cdot \frac{BC}{AB} \cdot \frac{AC}{BC} = 1.$$

By the converse of Ceva's Theorem, AD, BE, CF are concurrent. □

Later on, we will also re-derive the result on the concurrency of the altitudes of a triangle (see Theorem 3.1) using Ceva's Theorem; see page 87.

Exercise 3.23. Let D, E, F on BC, CA, AB in $\triangle ABC$ be such that

(1) starting from A, D divides the perimeter of $\triangle ABC$ into two equal parts,

(2) starting from B, E divides the perimeter of $\triangle ABC$ into two equal parts,

(3) starting from C, F divides the perimeter of $\triangle ABC$ into two equal parts.

Show that AD, BE, CF are concurrent.

3.3 Pythagoras's Theorem

In a right angled triangle, the side opposite to the right angle is called the _hypotenuse_.

Theorem 3.8 (Pythagoras's Theorem). _In a right angled triangle, the square of the hypotenuse is equal to the sum of the squares of the other two sides._

Proof. Let ABC be the right angled triangle with the right angle at the vertex A. Describe squares $ABB''A'$, $ACC'A''$, $BCC''B'$, having areas AB^2, AC^2, BC^2, respectively. Hence we reduce the problem to showing that the area of the big square is the sum of the areas of the two little squares. Draw a line through A, parallel to BB', meeting BC in D and $B'C''$ in D'.

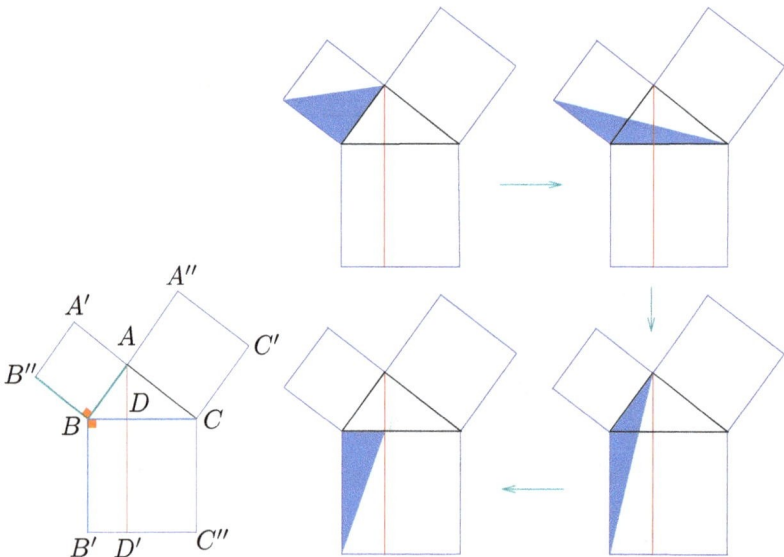

In the rightmost four pictures above, the shaded area is the same, because:

(1) $\triangle AB''B = \triangle CB''B$, because they share the same base $B''B$, and the vertices A, C lie on the line $A'C$ parallel to the base $B''B$.
(2) $\triangle B''BC = \triangle ABB'$, since these two triangles are congruent by the SAS Congruency Rule (indeed, we have $B''B = AB$, $BC = BB'$ and $\angle B''BC = 90° + \angle ABC = \angle ABB'$).

(3) $\triangle ABB' = \triangle DBB'$, since they share the same base BB', and the vertices A, D lie on the line AD' parallel to the base BB'.

Thus

$$\text{area of the square } ABB''A' = 2 \cdot \triangle AB''B = 2 \cdot \triangle BB'D$$
$$= \text{area of the rectangle } BB'D'D.$$

Similarly the area of the square $ACC'A''$ equals the area of the rectangle $CDD'C'''$.

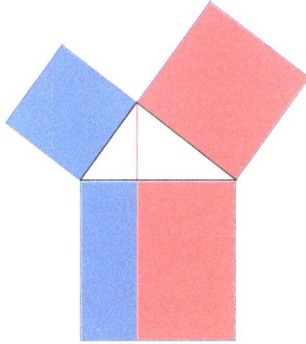

But the area of the square $BCC''B'$ is the sum of the areas of the two rectangles $BDD'B'$ and $CDD'C'''$. So the areas of the two little squares $ABB''A'$ and $ACC'A''$ add up to the area of the big square $BCC''B'$. In other words, we have $AB^2 + AC^2 = BC^2$. ☐

Exercise 3.24. Given a unit length in the plane, construct $\sqrt{2}$ using a straight edge and a compass. Can you construct $\sqrt{3}$? How about $\sqrt{n}$ for $n \in \mathbb{N}$?

Exercise 3.25. A rectangular box has length, width and height equal to 12 inches, 4 inches and 3 inches, respectively. What is the length of the diagonal of the box from a lower corner to the opposite upper corner of the box?

Exercise 3.26. A cylindrical cup has height 5 inches and radius of the circular base equal to $12/\pi$ inches. An ant sits on the rim, and wishes to walk along the curved surface of the cup to reach a sugar grain at exactly the opposite end on the base of the cup as shown in the picture. What is the length of the shortest path it can take?

Exercise 3.27. (Bhaskara[5]'s Proof of the Pythagoras Theorem). Consider a right angled triangle with length of the hypotenuse equal to c, and the lengths of the other two sides equal to a, b with $b > a$. If we take four copies of this triangle, and place their hypotenuses along the sides of a square, then show that we will obtain the following configuration, where the inner quadrilateral (shown shaded) is a square with side length $b - a$, and deduce Pythagoras's Theorem from this.

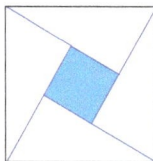

Exercise 3.28. (Pythagorean triples). A *Pythagorean triple* is a triple (a, b, c) of natural numbers a, b, c such that $c^2 = a^2 + b^2$.

(1) Check that $(3, 4, 5), (5, 12, 13)$ are Pythagorean triples.

(2) Prove that if (a, b, c) is a Pythagorean triple and if $n \in \mathbb{N}$, then (na, nb, nc) is also a Pythagorean triple. So, while studying Pythagorean triples (a, b, c), we can assume that a, b, c are coprime[6] a, b, c. Such Pythagorean triples are said to be *primitive*.

(3) (∗) Pythagorean triples have many interesting number theoretic properties. For example, show that if (a, b, c) is a Pythagorean triple, then $(c-b)(c-a)/2$ is a perfect square by considering the following picture.

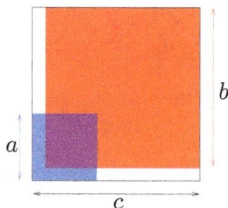

(4) (∗) Show that in a Pythagorean triple, 3 divides one of the numbers.

(5) Show that $(2nm, n^2 - m^2, n^2 + m^2)$, where $n, m \in \mathbb{N}$ and $n > m$, are Pythagorean triples. Thus we see that there are infinitely many Pythagorean triples[7].

---

[5]12th century Indian Mathematician.

[6]Recall that the *greatest common divisor* of a collection of natural numbers is a natural number such that it divides each of the numbers, and is divisible by each common divisor of the given numbers. A bunch of numbers are called *coprime* if their greatest common divisor is 1.

[7]This is in striking contrast with Fermat's Last Theorem, which says that there are no integer solutions to the equation $x^n + y^n = z^n$ for integer $n > 2$. This statement was

Exercise 3.29. (∗) Can a right angled triangle with integer sides have the non-hypotenuse sides equal to twin primes? (Primes p, p' are called *twin primes* if $|p - p'| = 2$. For example, $11, 13$ are twin primes.)

Exercise 3.30. (A physical proof of Pythagoras's Theorem). Imagine a fish tank with a right-triangular base, mounted horizontally on a vertical axis as shown in the picture. As a perpetual motion machine cannot exist, the rotational moment about the vertical axis must add to zero. Show that this yields Pythagoras's Theorem!

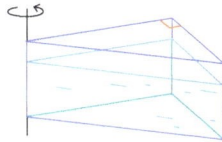

Exercise 3.31. Show that each altitude in an equilateral triangle of side length a has length $\sqrt{3}a/2$, and that the area of the triangle is $\sqrt{3}a^2/4$.

Exercise 3.32. (∗) An *integer lattice* or a *square lattice* in the plane is made as follows. Start by choosing an arbitrary point in the plane and an arbitrary length. Mark off all points in the plane which are obtained by a vertical displacement of that point through an integer multiple of the chosen length. Move each of the new points horizontally through integer multiples of the chosen length. This gives a square/integer lattice of points as shown in the leftmost picture below. Note that if we choose any point A in a square lattice, and rotate another lattice point B about the point A by $90°$, then we get a new lattice point B'.

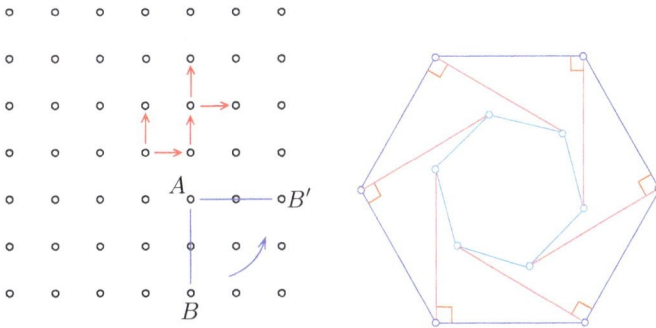

By considering the rightmost picture above, show that it is impossible to have a regular hexagon in the plane, all of whose vertices lie on a square lattice.

mentioned by Fermat in 1637 in the margin of a copy of *Arithmetica* where he claimed he had a proof, but that it was too large to fit in the margin. After over 300 years, a first successful proof was given by Andrew Wiles in 1995. Fermat himself gave a proof for the case $n = 4$, using an elementary number theoretic argument. The interested reader is referred to [Dolan(2011)] for a proof of this case.

Exercise 3.33. (Relativistic time dilation). Suppose that a rod AB moves with a constant speed v with respect to an observer O in a direction perpendicular to its length. Point A emits a light signal which is received at point B after a time t_0 measured in the reference frame moving with the rod, and at a time t measured by the observer O. Thus the rod moves to $A'B'$, at a distance vt from AB, as seen by O. We have $AB = ct_0$, $AB' = ct$ and $BB' = vt$.

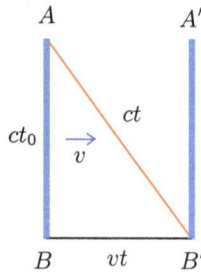

Using Pythagoras's Theorem in the right triangle $\triangle ABC$, derive the Time Dilation Formula in Special Relativity:

$$t = \frac{t_0}{\sqrt{1 - \dfrac{v^2}{c^2}}}.$$

Exercise 3.34. In the mathematical treatise *Lilavati* written by the 12th century Indian Mathematician Bhaskara, the following problem is posed:

A snake's hole is at the foot of a pillar, 9 "hasta" (= cubit = the length from the elbow to the tip of the middle finger) high, and a peacock is perched on its summit. Seeing a snake, at a distance of thrice the pillar height, gliding towards his hole, he pounces obliquely upon him. Say quickly at how many cubits from the snake's hole they meet, both proceeding an equal distance.

Exercise 3.35. [8] (Upside down Pythagoras). Show that in a right angled triangle, if the distance of the right angled vertex to the other two vertices is a, b, and its distance to the hypotenuse is d, then

$$\frac{1}{a^2} + \frac{1}{b^2} = \frac{1}{d^2}.$$

Exercise 3.36. (Converse of Pythagoras's Theorem). If $AB^2 + AC^2 = BC^2$ in a triangle $\triangle ABC$, then show that $\angle A = 90°$.

---

[8]This exercise can wait until similarity (Chapter 4) has been discussed.

Theorem 3.9 (Apollonius's Theorem). [9] *In a triangle ABC, if AD is the median from A meeting the opposite side BC in D, then*

$$AB^2 + AC^2 - \frac{BC^2}{2} = 2 \cdot AD^2.$$

Proof. Among the complementary angles $\angle ADB$ and $\angle ADC$, suppose that $\angle ADC$ is obtuse or right angled. (Otherwise just swap the labels B, C, and note that the equation above doesn't change!) Drop a perpendicular from A to the side BC, meeting BC in X, which lies outside DC, thanks to our assumption that $\angle ADC$ is obtuse.

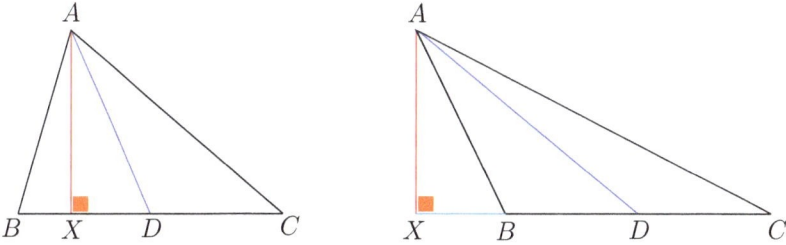

In each of the pictures above, we have by Pythagoras's Theorem applied to the triangles $\triangle AXB$, $\triangle AXD$ and $\triangle AXC$, that

$$AB^2 = AX^2 + BX^2 = (AD^2 - XD^2) + BX^2,$$
$$AC^2 = AX^2 + XC^2 = (AD^2 - XD^2) + XC^2.$$

Adding these, we obtain $AB^2 + AC^2 = 2 \cdot AD^2 + (BX^2 + XC^2 - 2 \cdot XD^2)$. We will be done if we prove that the expression in brackets above is $BC^2/2$. Note that in the left-hand side picture above we have $BX = BD - XD$, while in the right-hand side picture we have $BX = XD - BD$. But in either case, $BX = |BD - XD|$. Also in either of the two pictures, we have $XC = XD + DC$. Hence

$$BX^2 = |BD - XD|^2 = (BD - XD)^2 = BD^2 - 2 \cdot BD \cdot XD + XD^2$$
$$XC^2 = (XD + DC)^2 = DC^2 + 2 \cdot DC \cdot XD + XD^2.$$

But as D is the midpoint of BC, $BD = DC = BC/2$, and so upon adding the above,

$$BX^2 + XC^2 = BD^2 - 2 \cdot \cancel{BD \cdot XD} + XD^2 + DC^2 + 2 \cdot \cancel{DC \cdot XD} + XD^2$$
$$= \frac{BC^2}{2} + 2 \cdot XD^2.$$

This completes the proof. □

---

[9] Named after the Greek mathematician, Apollonius, c.262-190 BC.

Exercise 3.37. If G is the centroid of a triangle $\triangle ABC$, then show that

$$AB^2 + BC^2 + CA^2 = 3(GA^2 + GB^2 + GC^2).$$

Theorem 3.10. *If ℓ is a line and O is a point not on ℓ, then the point P on ℓ closest to O is such that $OP \perp \ell$.*

Proof. Drop a perpendicular from O to the line ℓ meeting ℓ at P. (Construction 2.6.)

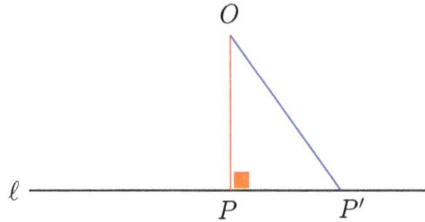

Then if P' is any other point on ℓ, by Pythagoras's Theorem applied to the right angled triangle $\triangle OPP'$, we have

$$OP' = \sqrt{OP^2 + PP'^2} > \sqrt{OP^2 + 0^2} = OP.$$

$\square$

Theorem 3.11 (Heron's Formula). [10]
The area of a triangle with sides a, b, c is

$$\sqrt{s(s-a)(s-b)(s-c)},$$

where $s := \dfrac{a+b+c}{2}$ is the semiperimeter of the triangle.

Proof. We'd like to use the following figure where the foot of the altitude dropped from C lies between the endpoints A and B of the side AB opposite to C. We ensure this as follows. In any triangle, there can be at most one angle with is obtuse. As the area expression is symmetric in a, b, c, there is no loss of generality in assuming that the obtuse angle in $\triangle ABC$, if there is one at all, is at C. Irrespective of whether the angle at C is acute or obtuse, we know then that the angles at A and B are acute. Then if we drop an altitude from C to the side AB, we know that it meets the opposite side AB at a point F *between* A and B.

---
[10]Named after the Greek Mathematician Hero of Alexandria, c.10-70 AD.

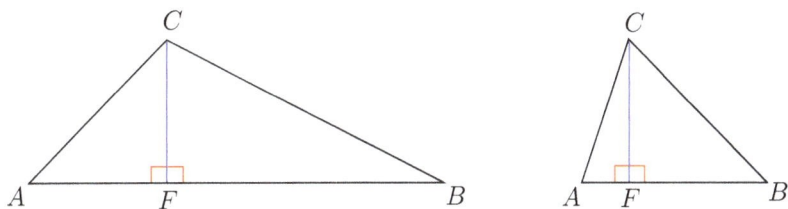

By Pythagoras's Theorem applied to the triangles $\triangle ACF$ and $\triangle CFB$, we have

$$CF^2 = b^2 - AF^2 = a^2 - (c - AF)^2,$$

so that

$$b^2 - \cancel{AF^2} = a^2 - c^2 + 2c \cdot AF - \cancel{AF^2},$$

that is,

$$AF = \frac{b^2 + c^2 - a^2}{2c}.$$

Hence the altitude CF is given by

$$CF = \sqrt{b^2 - AF^2} = \sqrt{b^2 - \left(\frac{b^2 + c^2 - a^2}{2c}\right)^2}$$

$$= \frac{\sqrt{4b^2c^2 - (b^2 + c^2 - a^2)^2}}{2c} = \frac{\sqrt{(2bc)^2 - (b^2 + c^2 - a^2)^2}}{2c}$$

$$= \frac{\sqrt{(2bc - (b^2 + c^2 - a^2))(2bc + (b^2 + c^2 - a^2))}}{2c}$$

$$= \frac{\sqrt{(a^2 - (b^2 + c^2 - 2bc))((b^2 + c^2 + 2bc) - a^2)}}{2c}$$

$$= \frac{\sqrt{(a^2 - (b - c)^2)((b + c)^2 - a^2)}}{2c}$$

$$= \frac{\sqrt{(a - (b - c))(a + (b - c))((b + c) - a)((b + c) + a)}}{2c}$$

$$= \frac{\sqrt{(a + b + c)(b + c - a)(a + c - b)(a + b - c)}}{2c}$$

$$= \frac{\sqrt{2s \cdot 2(s - a) \cdot 2(s - b) \cdot 2(s - c)}}{2c} = 2\frac{\sqrt{s(s - a)(s - b)(s - c)}}{c}.$$

So the area of $\triangle ABC$ is $\frac{1}{2}CF \cdot c = \sqrt{s(s - a)(s - b)(s - c)}.$ □

Exercise 3.38. (A Pythagoras type result). Points O, X, Y, Z in space are such that OX, OY, OZ are mutually perpendicular. Show that

$$(\triangle XYZ)^2 = (\triangle OXY)^2 + (\triangle OYZ)^2 + (\triangle OZX)^2.$$

Notes

The derivation of Kepler's Second Law is based on [Tong (2012)].
Exercise 3.30 is based on [Tokieda (1998)].

Chapter 4

Similar triangles

Two figures in the plane are congruent if by a rigid transformation (reflection, translation or rotation), one figure can be transformed into the other. Thus congruent triangles have the same shape and size. But now we will look at a more general type of transformation which also allows "scaling". Imagine looking at a portion of the plane through a magnifying glass for example. The image you see has the same shape as what is actually there on the page, except that its size has been "scaled" (magnified). Then we say that the resulting image is *similar* to the original image. For example, in the pictures below, we see scaled versions of the square by factors of 2, $1/2$, $\sqrt{2}$, and each of the resulting squares is similar to each of the others.

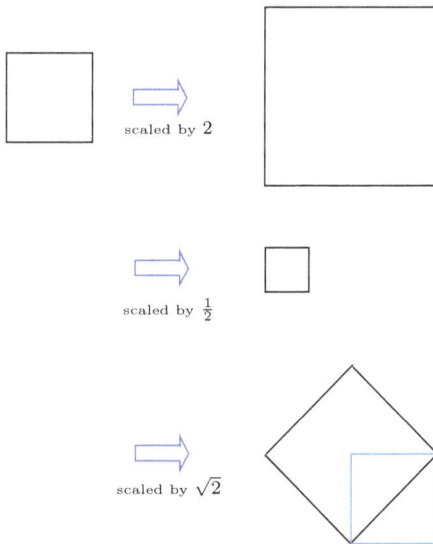

scaled by 2

scaled by $\frac{1}{2}$

scaled by $\sqrt{2}$

A similarity transformation preserves the shape, but not necessarily the size of the figure. Nevertheless, after a similarity transformation taking one figure to a new one, the new distances between any two pairs of points in the transformed figure are a fixed multiple of the distances between the corresponding pairs of points in the original figure. Thus, in the picture below, if A, B, C, D in the original picture correspond to the points A', B', C', D' in the new picture obtained after a similarity transformation, and if $C'D' = \lambda \cdot CD$, then $A'B' = \lambda \cdot AB$.

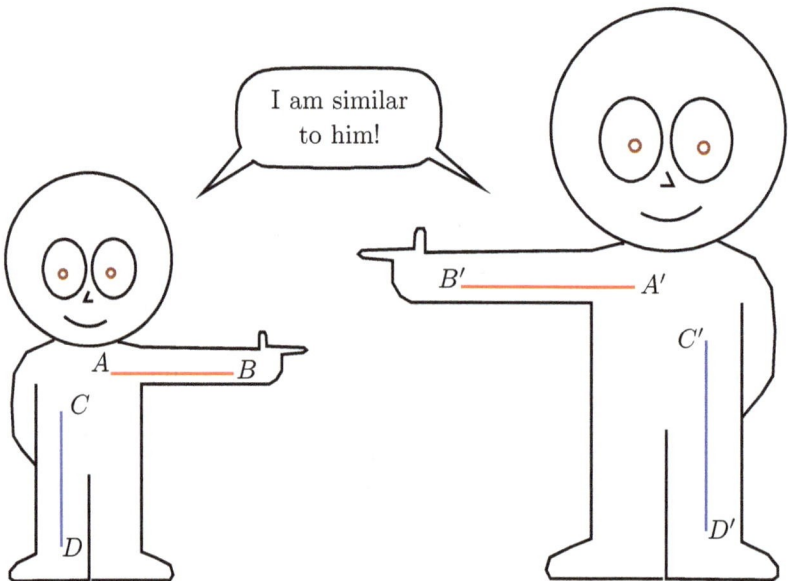

In this chapter, we will study similarity of triangles. Two triangles will be similar, if one can be obtained from the other by means of a similarity transformation. Since a similarity transformation preserves the shape and scales the size, it follows that the two triangles will be similar if their corresponding angles are equal (so that the shape is the same), and their corresponding sides are in the same ratio (so that one triangle is a scaled version of the other).

Definition 4.1 (Similar triangles). Two triangles are *similar* if their corresponding angles are equal and their corresponding sides are in the same ratio.

Notation. If $\triangle ABC$ is similar to $\triangle A'B'C'$ with

$$\angle A = \angle A', \quad \angle B = \angle B', \quad \angle C = \angle C',$$
$$\text{for some } k > 0, \quad AB = k \cdot A'B', \quad BC = k \cdot B'C', \quad CA = k \cdot C'A',$$

then we write $\triangle ABC \sim \triangle A'B'C'$.

Exercise 4.1. Show that all equilateral triangles are similar to each other.

Exercise 4.2. Show that similarity is an equivalence relation of the set of all triangles. (Similarity is of course a "weaker" equivalence relation than congruency, since a pair of congruent triangles are certainly also similar to each other, but a pair of similar triangles may not be congruent to each other.)

4.1 Basic Proportionality Theorem

Theorem 4.1 (Basic Proportionality Theorem). *In a triangle, a line drawn parallel to a side divides the other two sides in the same ratio.*

Proof. Let a line parallel to the side BC in the triangle $\triangle ABC$ meet the sides AB, AC at the points D, E, respectively. Draw the altitudes DX and EY in $\triangle ADE$.

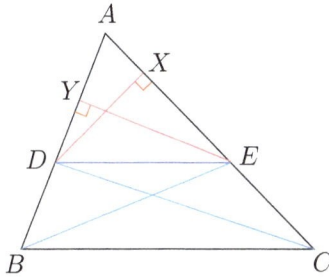

Then the area $\triangle ADE = \dfrac{AE \cdot DX}{2} = \dfrac{AD \cdot EY}{2}$, and so

$$\frac{AD}{AE} = \frac{DX}{EY}. \tag{4.1}$$

The triangles $\triangle BDE$ and $\triangle CDE$ share the same base (DE) and the vertices B, C lie on a line parallel to this base DE. Consequently their areas are equal, giving $DB \cdot EY = 2\triangle BDE = 2\triangle CDE = EC \cdot DX$. Hence

$$\frac{DB}{EC} = \frac{DX}{EY}. \tag{4.2}$$

From (4.1) and (4.2), we obtain $\dfrac{AD}{AE} = \dfrac{DB}{EC}$, and so $\dfrac{AD}{DB} = \dfrac{AE}{EC}$. $\qquad\square$

Proposition 4.1. *In a triangle* $\triangle ABC$, *let* D, E *be points on the sides* AB *and* AC. *Then the following are equivalent:*

(1) $\dfrac{AD}{DB} = \dfrac{AE}{EC}$

(2) $\dfrac{DB}{AB} = \dfrac{EC}{AC}$

(3) $\dfrac{AD}{AB} = \dfrac{AE}{AC}.$

Proof. We will show that $(1) \Rightarrow (2) \Rightarrow (3) \Rightarrow (1)$.

$(1) \Rightarrow (2)$: Adding 1 to both sides of (1) gives

$$\frac{AB}{DB} = \frac{AD + DB}{DB} = \frac{AD}{DB} + 1 = \frac{AE}{EC} + 1 = \frac{AE + EC}{EC} = \frac{AC}{EC}.$$

Taking reciprocals, we obtain (2).

$(2) \Rightarrow (3)$: Subtracting 1 to both sides of (2) gives

$$-\frac{AD}{AB} = \frac{DB - AB}{AB} = \frac{DB}{AB} - 1 = \frac{EC}{AC} - 1 = \frac{EC - AC}{AC} = -\frac{AE}{AC}.$$

Multiplying both sides by -1 gives (3).

$(3) \Rightarrow (1)$: Taking reciprocals in (3) gives $\dfrac{AB}{AD} = \dfrac{AC}{AE}$, and upon subtracting 1 from both sides gives

$$\frac{DB}{AD} = \frac{AB - AD}{AD} = \frac{AB}{AD} - 1 = \frac{AC}{AE} - 1 = \frac{AC - AE}{AE} = \frac{EC}{AE}.$$

Taking recipricals again, we obtain (1).   □

A consequence of this proposition and the Basic Proportionality Theorem is the following result.

Corollary 4.1. *In a triangle* $\triangle ABC$ *if points* D, E *on the sides* $AB, AC,$ *respectively are such that* $DE \parallel BC$, *then*

$$\frac{DB}{AB} = \frac{EC}{AC} \quad and \quad \frac{AD}{AB} = \frac{AE}{AC}.$$

Let us now prove the converse of the Basic Proportionality Theorem.

Theorem 4.2 (Basic Proportionality Theorem 2). *If a line divides two sides of a triangle in the same ratio, then it must be parallel to the third side.*

Proof. Suppose that a line intersects the sides AB, AC of a triangle $\triangle ABC$ in the points D, E, respectively, such that

$$\frac{AD}{DB} = \frac{AE}{EC}. \tag{4.3}$$

We want to show that $DE \parallel BC$. Let us draw a line through D parallel to BC, which meets the side AC at some point E'.

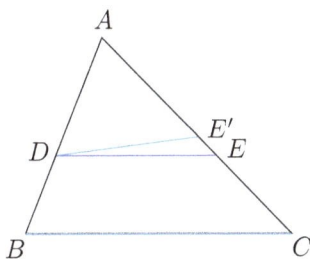

As $DE' \parallel BC$, it follows by the Basic Proportionality Theorem that

$$\frac{AD}{DB} = \frac{AE'}{E'C}. \tag{4.4}$$

From the equations (4.3) and (4.4), we obtain $\dfrac{AE}{EC} = \dfrac{AE'}{E'C}$.

Adding one to both sides gives

$$\frac{AC}{EC} = \frac{AC}{E'C},$$

and so $EC = E'C$, that is, E coincides with E'. Hence $DE \parallel BC$. □

Exercise 4.3. In the picture below, the diagonals of a quadrilateral $ABCD$ intersect each other at a point P. Show that $ABCD$ is a trapezium with $AB \parallel CD$ if and only if

$$\frac{AP}{PC} = \frac{BP}{PD}.$$

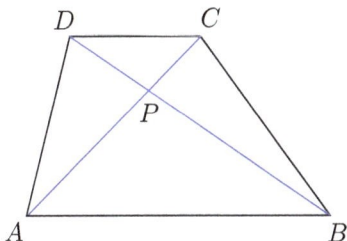

4.2   Criteria for similarity of triangles

The AAA Similarity Rule

Theorem 4.3 (AAA Similarity Rule). *If the three angles of one trian-gle are equal to the three angles of another triangle, then the two triangles are similar.*

Proof. Suppose that in two triangles $\triangle ABC$ and $\triangle A'B'C'$,

$$\angle A = \angle A', \quad \angle B = \angle B', \quad \angle C = \angle C'.$$

We would like to show that $\triangle ABC \sim \triangle A'B'C'$. In order to show that the two triangles are similar, we need to show that the corresponding angles are equal (already given) and that the sides are in the same ratio (and this is what we need to show). To this end, suppose that we mark a point P along the ray $\overrightarrow{A'B'}$ such that $A'P = AB$. Similarly, mark a point Q along the ray $\overrightarrow{A'C'}$ such that $A'Q = AC$.

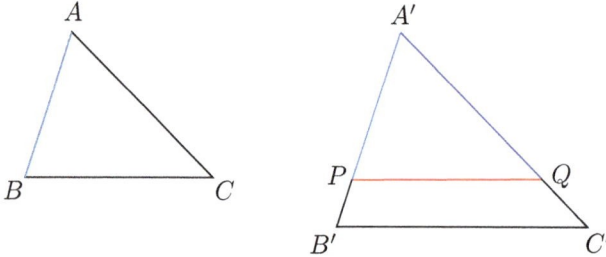

Then by the SAS Congruency Rule, $\triangle BAC \simeq \triangle PA'Q$. So $\angle A'PQ = \angle B$ (CPCT). But we know that $\angle B = \angle B'$ (given). Hence $\angle A'PQ = \angle B'$. This means that $PQ \parallel B'C'$. By the Basic Proportionality Theorem, it follows that

$$\frac{A'P}{A'B'} = \frac{A'Q}{A'C'}.$$

But by construction $A'P = AB$ and $A'Q = AC$. So the above gives

$$\frac{AB}{A'B'} = \frac{AC}{A'C'}.$$

In a similar manner, we can also show that

$$\frac{AC}{A'C'} = \frac{BC}{B'C'}.$$

Thus $\dfrac{AB}{A'B'} = \dfrac{BC}{B'C'} = \dfrac{AC}{A'C'}$. Consequently $\triangle ABC \sim \triangle A'B'C'$.   □

Corollary 4.2 (AA Similarity Rule). *If two angles of a triangle are equal to the two angles of another triangle, then they are similar.*

Proof. Since the sum of the angles in each triangle is 180°, also the third angle in one triangle is equal to the third angle of the other triangle. □

Here's another proof of the concurrency of the altitudes in a triangle (Theorem 3.1 on page 56).

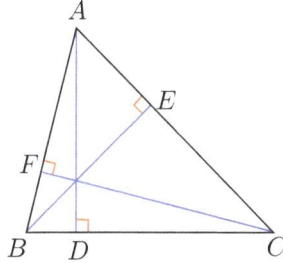

Proof. (A different proof of Theorem 3.1.)
In $\triangle AFC$ and $\triangle AEB$, $\angle A$ is common to both, $\angle AFC = \angle AEB = 90°$.
By the AA Similarity Rule, $\triangle AFC \sim \triangle AEB$. Hence $\dfrac{AF}{EA} = \dfrac{AC}{AB}$.
Similarly, $\dfrac{BD}{FB} = \dfrac{AB}{BC}$ and $\dfrac{CE}{DC} = \dfrac{BC}{AC}$.
Thus $\dfrac{BD}{DC} \cdot \dfrac{CE}{EA} \cdot \dfrac{AF}{FB} = \dfrac{BD}{FB} \cdot \dfrac{CE}{DC} \cdot \dfrac{AF}{EA} = \dfrac{AB}{BC} \cdot \dfrac{BC}{AC} \cdot \dfrac{AC}{AB} = 1.$
By the converse of Ceva's Theorem, AD, BE, CF are concurrent. □

Exercise 4.4. (∗) A piece of a map indicating a treasure is found, in which the treasure is known to lie on an island in the sea at the point of intersection of two nonparallel lines on the map. Unfortunately, the piece of map found does not contain the very point of intersection. The captain of a ship realizes that his ship lies at a position P on the map between the two lines.

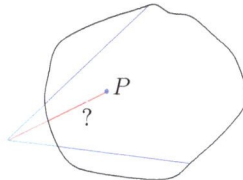

In order to determine the direction in which he should sail, he would like to construct a line passing through P which will meet the point of intersection of the two lines if it were extended. How would we perform this construction on the piece of map using a straight edge and a compass?

Exercise 4.5. (∗) Bisect a line segment with a straight edge alone, given only a line parallel to it.

Sum of the Geometric Series

Given $r > 0$, the sequence of numbers $a, ar, ar^2, ar^3, \cdots$ is called a *geometric progression* with *common ratio* r. Each successive term is obtained from the previous one by multiplication by r.

For example, think of a ball bouncing on the floor. Assume that the collision is inelastic with a "coefficient of restitution" (that is, the ratio of the speed after the collision to the one before) equal to $k < 1$. Then the height of the ball reduces after each bounce, and the resulting sequence of heights can be shown to be in geometric progression. Indeed, if the instantaneous speed at the ground is v, then the height reached is given by $v^2/(2g)$, where g is the acceleration due to gravity, and so if the initial height is H, then the subsequent heights are $H, k^2 H, k^4 H, k^6 H, \cdots$, forming a geometric progression with common ration k^2. One can ask: What is the total vertical distance travelled by such a bouncing ball? That is:

$$H + 2k^2 H + 2k^4 H + 2k^6 H + \cdots = H + 2H(1 + k^2 + (k^2)^2 + (k^2)^3 + \cdots) = ?$$

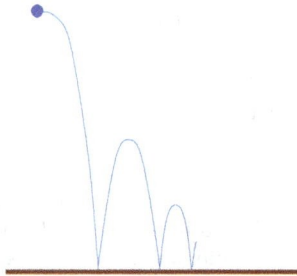

More generally, we would like to "sum the *geometric series*"

$$a + ar + ar^2 + ar^3 + \cdots =: a \sum_{n=0}^{\infty} r^n.$$

Let $0 < r < 1$. We'll use geometry to find the sum $1 + r + r^2 + r^3 + \cdots$. Let

$$S = 1 + r + r^2 + r^3 + \cdots.$$

Consider a right angled triangle $\triangle ABC$ with nonhypotenuse side lengths $AB = 1$ and $BC = S$ as shown. Mark a point B' on BC so that $BB' = 1$.

Then $B'C = S - 1 = r + r^2 + r^3 + r^4 + \cdots = r(1 + r + r^2 + r^3 + \cdots) = r \cdot S$. Erect a perpendicular at B', meeting AC in A'. As $\triangle ABC$ and $\triangle A'B'C'$ are similar, we have

$$\frac{A'B'}{1} = \frac{A'B'}{AB} = \frac{B'C}{BC} = \frac{r \cdot S}{S} = r.$$

Now draw a line through A' meeting AB in D. Then

$$AD = AB - DB = AB - A'B' = 1 - r.$$

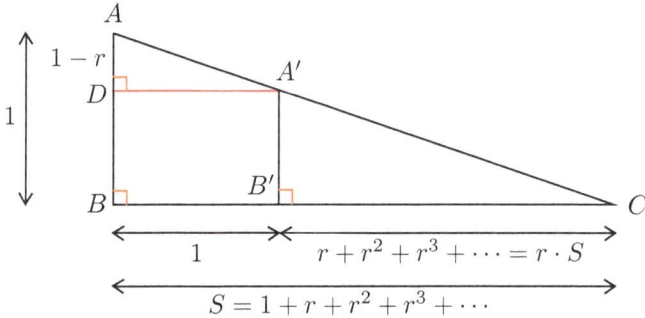

Since the two triangles $\triangle ADA'$ and $\triangle ABC$ are similar, it follows that

$$\frac{S}{1} = \frac{BC}{BB'} = \frac{BC}{DA'} = \frac{AB}{AD} = \frac{1}{1 - r}.$$

Thus

$$a \sum_{n=0}^{\infty} r^n = \frac{a}{1 - r}.$$

Exercise 4.6. Using the pictures below, convince yourself of the facts that

$$\sum_{n=1}^{\infty} \left(\frac{1}{2}\right)^n = 1 \quad \text{and} \quad \sum_{n=1}^{\infty} \left(\frac{1}{2}\right)^{2n} = \sum_{n=1}^{\infty} \left(\frac{1}{4}\right)^n = \frac{1}{3}.$$

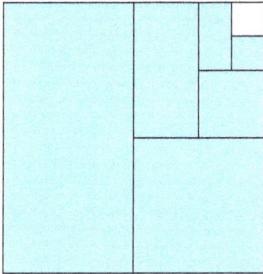

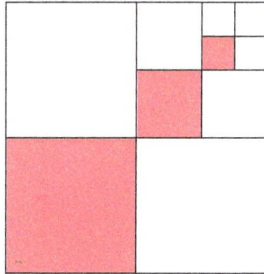

Recall the Angle Bisector Theorem (Theorem 3.5) which says that the (internal) angle bisector in a triangle divides the opposite side in the ratio of the sides bounding the angle. Now we will learn the validity of the conclusion also for *external* angle bisectors.

Theorem 4.4 (External Angle Bisector Theorem).
In a triangle $\triangle ABC$, suppose that the side BA is extended to a point A', and that the angle bisector of the exterior angle $\angle CAA'$ meets the side BC (extended) at the point D. Then $AB : AC = BD : CD$.

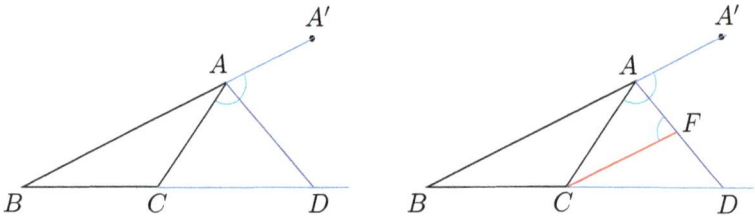

Proof. Draw a line through C, parallel to AB, meeting AD in F. Then we have that $\angle CAF = \angle FAA' = \angle AFC$, so that $\triangle ACF$ is isosceles, and $CF = AC$. But by the AA Similarity Rule, $\triangle DCF \sim \triangle DBA$, and so
$$\frac{BD}{CD} = \frac{AB}{CF} = \frac{AB}{AC}.$$
$\square$

Construction 4.1 (Division of a line segment into n equal parts).
Given a line segment, divide it into $n \in \mathbb{N}$ equal parts.

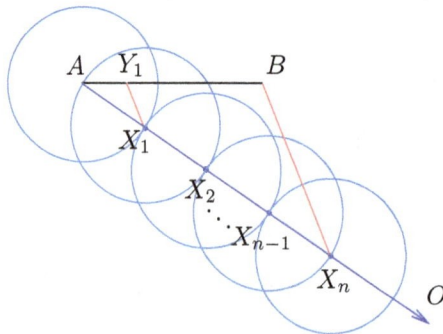

Draw a ray $\overrightarrow{AO}$ which intersects the line containing AB only at A. Mark off an arbitrary point X_1 on $\overrightarrow{AO}$, and subsequently, mark off points $X_2, \cdots, X_n$

such that the length of each $X_i X_{i+1}$ equals AX_1. Thus we have that $AX_1 = X_1 X_2 = \cdots = X_{n-1} X_n$. Hence

$$AX_1 = \frac{1}{n} AX_n.$$

Draw a line through X_1 parallel to BX_n, and suppose it meets AB in Y_1. Then we have $\triangle AY_1 X_1 \sim \triangle ABX_n$. Consequently,

$$AY_1 = \frac{AX_1}{AX_n} \cdot AB = \frac{1}{n} AB. \qquad \diamond$$

The number line. In elementary school, we learn about

the natural numbers $\mathbb{N} := \{1, 2, 3, \cdots\}$

the integers $\mathbb{Z} := \{\cdots, -3, -2, -1, 0, 1, 2, 3, \cdots\}$, and

the rational numbers $\mathbb{Q} := \left\{ \left[\frac{n}{d}\right] : n, d \in \mathbb{Z}, d \neq 0 \right\}.$

Incidentally, the rationale behind denoting the rational numbers by $\mathbb{Q}$ is that it reminds us of "**quotient**". The notation $\mathbb{Z}$ for integers stems from the German word "**zählen**" (meaning "count"). In the above,

$$\left[\frac{n}{d}\right]$$

represents a whole family of "equivalent fractions"; for example

$$\frac{2}{4} = \frac{1}{2} = \frac{-3}{-6} \text{ etc.}$$

We can visualize these numbers on the "number line". What is the number line? It is any line in the plane, on which we have chosen a point O as the "origin", representing the number 0, and chosen a unit length by marking off a point on the right of O, where the number 1 is placed. In this way, we get all the positive integers, $1, 2, 3, 4, \cdots$ by repeatedly marking off successively the unit length towards the right, and all the negative integers $-1, -2, -3, \cdots$ by repeatedly marking off successively the unit length towards the left.

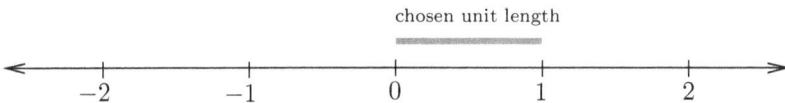

Just like the integers can be depicted on the number line, we can also depict all rational numbers on it, as follows. First of all, we know a procedure for dividing a unit length on the number line into d ($\in \mathbb{N}$) equal parts, allowing us to construct the rational number $1/d$ on the number line. Having obtained $1/d$, we can now construct n/d on the number line for *any* $n \in \mathbb{Z}$, by repeating the length $1/d$ n times towards the right of 0 if $n > 0$, and towards the left $-n$ times from 0 if n is negative.

Hence we can depict all the rational numbers on the number line. Does this exhaust the number line? That is, suppose that we start with all the points on the number line being coloured black, and suppose that at a later time, we colour all the rational ones by red: are there any black points left over? The answer is yes, and we will demonstrate this later. We will show that there does "exist", based on geometric reasoning, a point on the number line, whose square is 2, but we will also argue that this number, denoted by $\sqrt{2}$, is not a rational number.

First of all, the picture below shows that $\sqrt{2}$ exists as a point on the number line. Indeed, by looking at the right angled triangle $\triangle OBA$, Pythagoras's Theorem tells us that the length of the hypotenuse OA satisfies $OA^2 = OB^2 + AB^2 = 1^2 + 1^2 = 2$, and so OA is a number, denoted say by $\sqrt{2}$, whose square is 2. By taking O as the center and radius OA, we can draw a circle using a compass which intersects the number line at a point on the right of O, and this point corresponds to the number $\sqrt{2}$. This was done in Exercise 3.24. Is $\sqrt{2}$ a rational number? We'll show later on that it isn't!

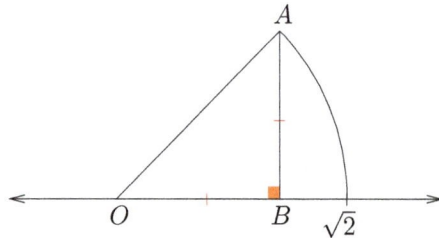

Thus we have seen that the elements of $\mathbb{Q}$ can be depicted on the number line, and that not all the points on the number line belong to $\mathbb{Q}$. We think of $\mathbb{R}$ as *all* the points on the number line. As mentioned before, if we take out everything on the number line (the black points) except for the rational numbers $\mathbb{Q}$ (the red points), then there will be holes amongst the rational numbers (for example there will be a missing black point where $\sqrt{2}$ lies on the number line. We can think of the real numbers as "filling in" these holes between the rational numbers.

Exercise 4.7. Depict $-11/6$ on the number line.

Construction 4.2 (Triangle from median lengths).

Given the lengths of the medians of a triangle, construct the triangle.

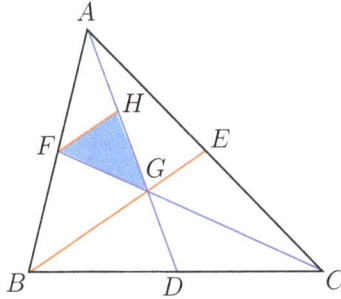

In the previous picture, AD, BE, CF are the medians, intersecting at the centroid G, and $FH \parallel BG$. Then using the Midpoint Theorem in $\triangle ABG$ and the fact that a centroid divides each median in the ratio $2 : 1$, we obtain

$$HG = \frac{1}{2}AG = \frac{1}{2} \cdot \frac{2}{3}AD = \frac{1}{3}AD,$$

$$FH = \frac{1}{2}BG = \frac{1}{2} \cdot \frac{2}{3}BE = \frac{1}{3}BE,$$

$$GF = \frac{1}{3}CF.$$

Thus knowing the lengths AD, BE, CF, and also a construction of trisection of a line segment, we know the side lengths of $\triangle FGH$, and so we can construct $\triangle FGH$. Then we can obtain the location of the points A, D by extending HG on either side such that $AH = HG$ and $DG = GH$. Then we join AF and extend it to B such that $BF = AF$. Join BD and extend it to C such that $BD = DC$. So we have obtained the location of the points A, B, C, and hence also $\triangle ABC$. ◇

Exercise 4.8. Prove the converse of the External Angle Bisector Theorem. That is, show that if D is a point on the extension of the side BC of $\triangle ABC$ such that $AB : AC = BD : CD$, then AD bisects the external angle at A.

Exercise 4.9.
(1) In $\triangle ABC$, M, N are points on the sides AC, BC, respectively such that $MN \parallel AB$. Suppose that AN and BM intersect in the point P. The line passing through C and P meets MN at L and AB in D. Prove that D and L are midpoints of the segments AB and MN respectively.

(2) Give a procedure for constructing the midpoint of a side of a square using a straight edge alone.

Exercise 4.10. In a quadrilateral $ABCD$, $AB \parallel CD$, and the diagonals AC, BD intersect at right angles. Show that $AD \cdot BC \geq AB \cdot CD$ and $AD + BC \geq AB + CD$. Prove that any of the inequalities becomes an equality if any only if $ABCD$ is a rhombus.

Exercise 4.11. (Estimation of the radius of the Earth). The Egyptian astronomer Erastosthenes (276-195 BC) estimated the radius of the Earth using similar triangles as follows. He observed that the length of a shadow cast by objects at noon at the same time of the year changed from place to place. On summer solstice, the spire in Alexandria cast a shadow about $\frac{1}{8}$th of its height, while in Syene, about 500 miles south of Alexandria, no shadow was cast at noon. Using the picture below, estimate Earth's radius $R_\oplus$ in miles and in kilometers à la Erastosthenes. See the picture below. Here $s/h \approx 1/8$ and $d \approx 500$ miles. Also, 1 mile is 1.6 km.

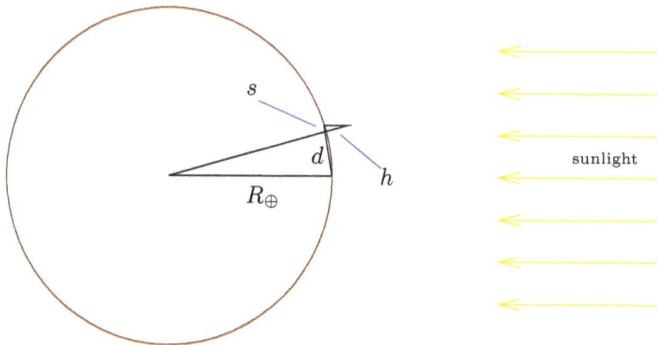

Theorem 4.5 (An "origami" proof of the irrationality of $\sqrt{2}$).
There is no rational number $q \in \mathbb{Q}$ such that $q^2 = 2$.

Proof. Suppose that $\sqrt{2}$ is a rational number. Then some scaling of the triangle

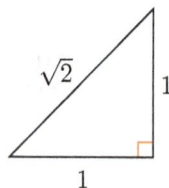

by an integer will produce a similar triangle, all of whose sides are integers. Choose the smallest such triangle, say $\triangle ABC$, with integer side lengths $BC = AB = n$, and $AC = N$, $n, N \in \mathbb{N}$. Now do the following origami: fold along a line passing through A so that B lies on AC, giving rise to the point B' on AC. The "crease" in the paper is actually the angle bisector AD of the angle $\angle BAC$.

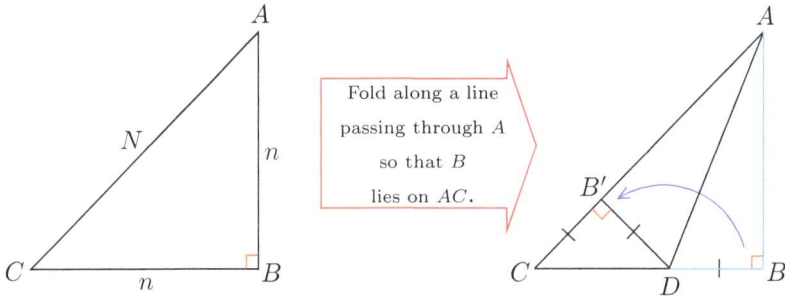

In $\triangle CB'D$, $\angle CB'D = 90°$, $\angle B'CD = 45°$. So $\triangle CB'D$ is an isosceles right triangle. We have $CB' = B'D = AC - AB' = N - n \in \mathbb{N}$, while

$$CD = CB - DB = n - B'C = n - (N - n) = 2n - N \in \mathbb{N}.$$

So $\triangle CB'D$ is similar to the triangle

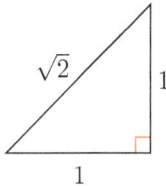

has integer side lengths, and is smaller than $\triangle ABC$, contradicting the choice of $\triangle ABC$. So there is no rational number q such that $q^2 = 2$. □

Here is a result akin to Ceva's Theorem. Recall that Ceva's Theorem dealt with *concurrency* of *lines*, that is, when three lines pass through a common point. Menelaus's Theorem, on the other hand, is about *collinearity* of *points*, that is, when three points lie on a line.

Theorem 4.6 (Menelaus). *If a line ℓ cuts (possibly extensions of) AB, BC, CA at the points F, D, E, then*

$$\frac{BD}{DC} \cdot \frac{CE}{EA} \cdot \frac{AF}{FB} = 1.$$

Proof. Draw a line through A parallel to BC, and suppose that the line ℓ passing through D, E, F meets this line at D'.

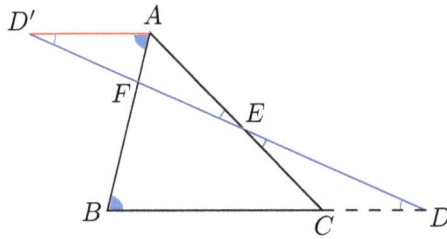

Then $\triangle AFD' \sim \triangle BFD$ by the AA Similarity Rule ($\angle AD'F = \angle BDF$ and $\angle D'AF = \angle DBF$).

So $\dfrac{AF}{FB} = \dfrac{AD'}{BD}$, that is,

$$AD' = \frac{AF \cdot BD}{FB}. \tag{4.5}$$

Also in triangles $\triangle AD'E$ and $\triangle CDE$, we have $\angle AD'E = \angle CDE$ and $\angle AED' = \angle CED$. Thus by the AA Similarity Rule, $\triangle AD'E \sim \triangle CDE$.

Hence $\dfrac{AD'}{DC} = \dfrac{EA}{CE}$, that is,

$$AD' = \frac{DC \cdot EA}{CE}. \tag{4.6}$$

From (4.5) and (4.6), $\dfrac{AF \cdot BD}{FB} = \dfrac{DC \cdot EA}{CE}$, and upon rearranging,

$$\frac{BD}{DC} \cdot \frac{CE}{EA} \cdot \frac{AF}{FB} = 1.$$

$\square$

Exercise 4.12. (Converse to Menelaus's Theorem). The converse of Menelaus's Theorem should say something like

"If the points D, E, F are on the lines containing the sides BC, CA, AB, respectively and

$$\frac{BD}{DC} \cdot \frac{CE}{EA} \cdot \frac{AF}{FB} = 1,$$

then D, E, F are collinear."

But we have to be careful, since we can't take the points D, E, F all in the interior of the sides of the triangle (for example, when D, E, F are the midpoints of the sides of the triangle), since in that case D, E, F can't be collinear, and the relevant result in this case is Ceva's Theorem (on concurrency). Show that the converse to Menelaus's Theorem holds when one of the points, say D lies outside BC, and the other two points are internal, that is, E lies inside CA, and F lies inside AB.

Exercise 4.13. A line ℓ cuts the (possibly extended) sides AB, BC, CD, DA of a quadrilateral $ABCD$ at the points P, Q, R, S, respectively. Prove that

$$\frac{AP}{PB} \cdot \frac{BQ}{QC} \cdot \frac{CR}{RD} \cdot \frac{DS}{SA} = 1.$$

Exercise 4.14. ($*$) Let's revisit Exercise 3.22. Recall that over there, in a triangle $\triangle ABC$, we had taken points C', A', B' on the sides AB, BC, CA, respectively, such that $AC' : C'B = BA' : A'C = CB' : BA = 2 : 1$. Join AA', BB', CC', and let the intersection points in the line segment pairs $(CC', AA'), (AA', BB'), (BB', CC')$ be labelled as X, Y, Z, respectively. What is the ratio of the areas of the internal triangle $\triangle XYZ$ to that of $\triangle ABC$? Next, suppose that $AC' : C'B = BA' : A'C = CB' : B'A = x > 0$. What is the ratio $\triangle XYZ : \triangle ABC$ then? Do you get the expected answer when $x = 0$ and when $x = 1$?

The SSS Similarity Rule

Theorem 4.7 (SSS Similarity Rule). *If in two triangles, the three ratios of the length of the side of one triangle to the other triangle are all the same, then the two triangles are similar.*

Proof. Let $\triangle ABC$, $\triangle A'B'C'$ be two triangles such that

$$\frac{AB}{A'B'} = \frac{AC}{A'C'} = \frac{BC}{B'C'}. \tag{4.7}$$

It is enough to consider the case when the above common ratio is < 1. (In case the common ratio is equal to 1, the corresponding sides are the same, so that by the SSS Congruency Rule the triangles are congruent, and hence also similar. When the common ratio is > 1, we simply swap the two triangles.) Mark a point P on $A'B'$ such that $A'P = AB$. Similarly, let the point Q on $A'C'$ be such that $A'Q = AC$.

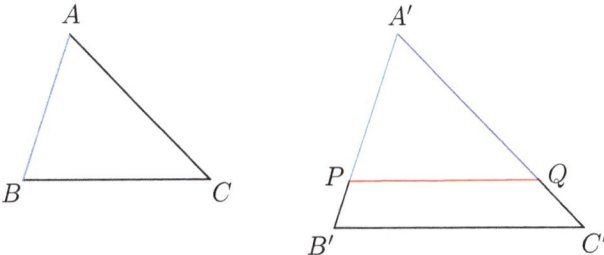

From (4.7), we obtain $\dfrac{A'B'}{A'P} = \dfrac{A'B'}{AB} = \dfrac{A'C'}{AC} = \dfrac{A'C'}{A'Q}$. Thus

$$\frac{PB'}{A'P} = \frac{A'B' - A'P}{A'P} = \frac{A'B'}{A'P} - 1 = \frac{A'C'}{A'Q} - 1 = \frac{A'C' - A'Q}{A'Q} = \frac{QC'}{A'Q}.$$

By the Basic Proportionality Theorem 2, it follows that PQ is parallel to $B'C'$. Thus we have $\angle A'PQ = \angle B'$ and $\angle A'QP = \angle C'$. By the AA Similarity Rule, $\triangle A'PQ \sim \triangle A'B'C'$. So the lengths of their respective sides must be proportional, and in particular, also using (4.7), we obtain

$$\frac{BC}{B'C'} = \frac{AB}{A'B'} = \frac{A'P}{A'B'} = \frac{PQ}{B'C'}.$$

Comparing the first and last terms above, $BC = PQ$. By the SSS Congruency Rule, $\triangle ABC \simeq \triangle A'PQ$. So $\angle A = \angle A'$ and $\angle B = \angle A'PQ \ (= \angle B')$. By the AA Similarity Rule, $\triangle ABC \sim \triangle A'B'C'$. □

The SAS Similarity Rule

Theorem 4.8 (SAS Similarity Rule). *If in two triangles, one angle of one triangle is equal to that of the other, and the sides including this angle have the same ratio of lengths, then the two triangles are similar.*

Proof. Let $\triangle ABC$, $\triangle A'B'C'$ be two triangles such that $\angle A = \angle A'$ and

$$\frac{AB}{A'B'} = \frac{AC}{A'C'}. \tag{4.8}$$

It is enough to consider the case when the above common ratio is < 1. (In case the common ratio is equal to 1, it can be seen by the SAS Congruency Rule that the triangles are congruent, and hence also similar. When the common ratio is > 1, we simply swap the two triangles.) Mark a point P on $A'B'$ such that $A'P = AB$. Similarly, let the point Q on $A'C'$ be such that $A'Q = AC$.

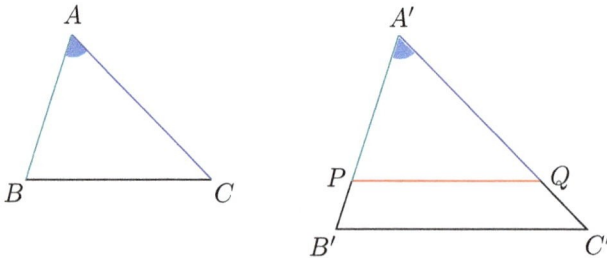

Then $\triangle ABC \simeq \triangle A'PQ$ by the SAS Congruency Rule, and so

$$\angle B = \angle A'PQ. \tag{4.9}$$

From (4.8), we obtain $\dfrac{A'B'}{A'P} = \dfrac{A'B'}{AB} = \dfrac{A'C'}{AC} = \dfrac{A'C'}{A'Q}$. Thus

$$\frac{PB'}{A'P} = \frac{A'B' - A'P}{A'P} = \frac{A'B'}{A'P} - 1 = \frac{A'C'}{A'Q} - 1 = \frac{A'C' - A'Q}{A'Q} = \frac{QC'}{A'Q}.$$

By the Basic Proportionality Theorem 2, it follows that PQ is parallel to $B'C'$. Thus we have $\angle A'PQ = \angle B'$. Using (4.9), $\angle B = \angle B'$. Since $\angle A = \angle A'$ (given) and $\angle B = \angle B'$ (just proved), we conclude using the AA Similarity Rule that $\triangle ABC \sim \triangle A'B'C'$. $\qquad\square$

Exercise 4.15.
(1) In the picture below, it is given that B is the midpoint of the line segment AA', and that $AA' = PA'$, $PA = PM = AB$. Show that M must be the midpoint of the side AB.

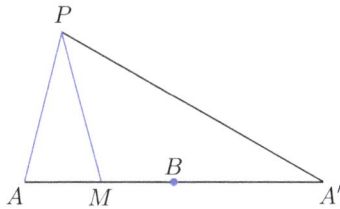

(2) ($*$) Given a line segment, give a method for constructing its midpoint using a compass alone.
(3) ($*$) Given a line segment and a natural number n, can you give a method for dividing it into n equal parts by using a compass alone?

Exercise 4.16. (Cauchy-Schwarz Inequality). By considering the picture below, show that for all positive a, b, c, d, $ab+cd \le \sqrt{a^2 + b^2}\sqrt{c^2 + d^2}$. Show that equality holds if and only if $a/c = b/d$.

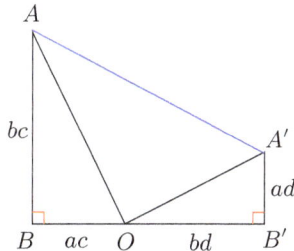

4.3 Areas of similar triangles

We have seen that similar triangles possess sides in the same ratio. What can we say about the ratio of their areas? We will now see that the areas of two similar triangles bear the same ratio as the *square* of the ratio of the lengths of their corresponding sides.

Theorem 4.9. *The ratio of the areas of two similar triangles is equal to the square of the ratio of their corresponding sides.*

Proof. A quick way to see this is by using Heron's Formula. Let the sides of the two triangles be a, b, c and ka, kb, kc (where k is the ratio of the lengths of their sides). If

$$s := \frac{a+b+c}{2}$$

is the semiperimeter of the first triangle, then the seimiperimeter of the second triangle is $k \cdot s$. Thus the ratio of the areas of the two triangles is

$$\frac{\sqrt{ks(ks-ka)(ks-kb)(ks-kc)}}{\sqrt{s(s-a)(s-b)(s-c)}} = k^2 \frac{\sqrt{s(s-a)(s-b)(s-c)}}{\sqrt{s(s-a)(s-b)(s-c)}} = k^2.$$

(Here is an alternative proof. Let $\triangle ABC \sim \triangle A'B'C'$. Then in particular we have

$$\frac{BC}{B'C'} = \frac{AB}{A'B'}. \tag{4.10}$$

Drop altitudes AD, $A'D'$ from the vertices A, A' in the two triangles. See the pictures below, for the two cases when B is acute (left set of pictures) and when B is obtuse (right set).

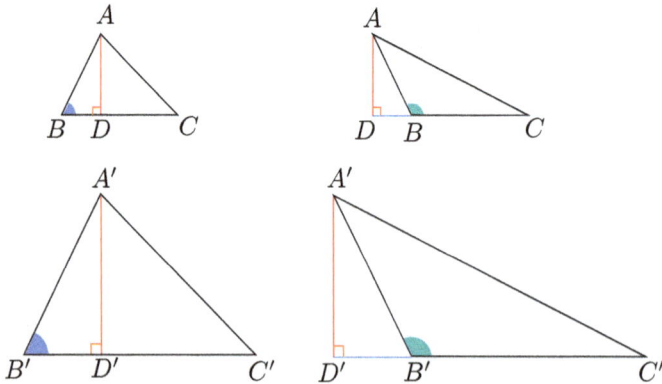

Then

$$\angle ABD = \angle ABC = \angle A'B'C' = \angle A'B'D' \text{ (acute case)},$$
$$\angle ABD = 180° - \angle ABC = 180° - \angle A'B'C' = \angle A'B'D' \text{ (obtuse case)}.$$

Also, $\angle ADB = \angle A'D'B' = 90°$. Thus by the AA Similarity Rule, $\triangle ADB \sim \triangle A'D'B'$. In particular,

$$\frac{AD}{A'D'} = \frac{AB}{A'B'}. \qquad (4.11)$$

Consequently, using (4.10) and (4.11)

$$\frac{\triangle ABC}{\triangle A'B'C'} = \frac{BC \cdot AD/2}{B'C' \cdot A'D'/2} = \frac{BC}{B'C'} \cdot \frac{AD}{A'D'} = \frac{AB}{A'B'} \cdot \frac{AB}{A'B'} = \left(\frac{AB}{A'B'}\right)^2.$$

This completes the alternative proof.) □

Exercise 4.17. (∗) Through a point inside a triangle, lines are drawn parallel to the three sides, resulting in three little triangles, as shown below. If the area of the given triangle is A, and the areas of the little triangles are A_1, A_2, A_3, then show that $\sqrt{A} = \sqrt{A_1} + \sqrt{A_2} + \sqrt{A_3}$.

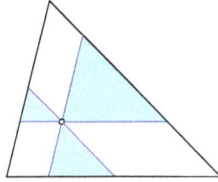

Exercise 4.18. Let $\triangle ABC$ be a right angled triangle, right angled at B, and let BE be the altitude dropped from B to the hypotenuse. Let r_1, r_2 be the inradii of the two smaller triangles formed, and let r be the inradius of $\triangle ABC$. Show that $r_1^2 + r_2^2 = r^2$.

We end this chapter with a rather sophisticated inequality due to the 20th century mathematicians Paul Erdős[1] and Louis Mordell[2]. Although the proof is rather involved, we have nevertheless included it since firstly, the result is "pretty" (and almost surprising that one has a factor of 2 appearing on the right-hand side of the following inequality), and secondly, to give a flavour of a 20th century result in Euclidean geometry. But the reader can feel free to skip this, and skipping it won't result in any loss of continuity in what follows.

---

[1] Paul Erdős (1913-96) was a very prolific Hungarian mathematician, who contributed to many diverse field in mathematics.
[2] Louis Mordell (1888-1972) was an American-born British mathematician.

Theorem 4.10 (Erdős-Mordell Inequality). *If P is a point inside a triangle $\triangle ABC$, and the line segments PL, PM, PN are the perpendiculars dropped from P to the sides, then*

$$PA + PB + PC \geq 2(PL + PM + PN).$$

If equality holds above, then $\triangle ABC$ is equilateral.

Proof. In the picture on the left below, we use

$$x, y, z \text{ for the lengths of } PA, PB, PC,$$
$$a, b, c \text{ for the lengths of } BC, CA, AB, \text{ and}$$
$$\ell, m, n \text{ for the lengths of } PL, PM, PN.$$

In the picture on the right below, we have scaled the original triangle $\triangle ABC$ by a factor of x to obtain the similar triangle $\triangle A_0 B_0 C_0$.

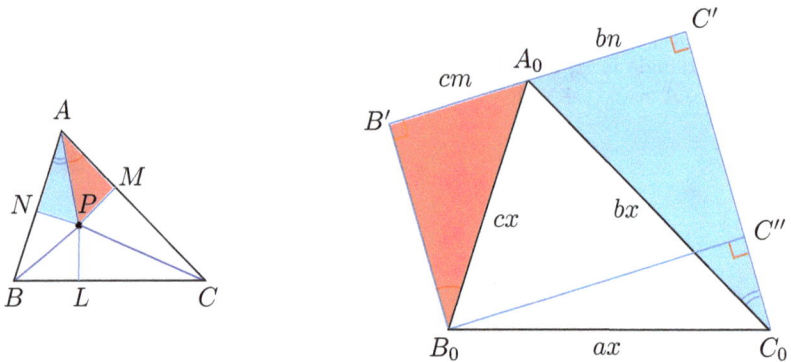

Hence $A_0 B_0 = cx$, $B_0 C_0 = ax$, $C_0 A_0 = bx$. Next we have marked off a point B' along a ray $\overrightarrow{B_0 B'}$ such that $\angle B' B_0 A_0 = \angle PAM$ and dropped a perpendicular from A_0 on this ray, meeting the ray in B'. Similarly, C' is a point such that $A_0 C'$ is perpendicular to $C_0 C'$ and $\angle A_0 C_0 C' = \angle PAN$. By the AA Similarity Rule, $\triangle PAN \sim \triangle A_0 C_0 C'$ (with scaling factor of side lengths equal to b) and $\triangle PAM \sim \triangle A_0 B_0 B'$ (with scaling factor c). Hence $A_0 B' = cm$ and $A_0 C' = bn$.

Next we will show that the three points B', A_0, C' are collinear. To this end, note that we have

$$\angle B' A_0 B_0 = 90° - \angle A_0 B_0 B' = 90° - \angle PAM,$$
$$\angle C' A_0 C_0 = 90° - \angle A_0 C_0 C' = 90° - \angle PAN, \text{ and}$$
$$\angle B_0 A_0 C_0 = \angle BAC = \angle PAM + \angle PAN.$$

Adding these, we obtain $\angle B'A_0C' = 180°$, so that the points B', A_0, C' are collinear, and $B_0B'C'C_0$ is a trapezium. So by dropping a perpendicular from B_0 to the side C_0C', meeting it in C'', we have $B_0C_0 \geq B_0C'' = B'C'$, that is,

$$ax \geq cm + bn.$$

Similarly, we also obtain

$$by \geq c\ell + an,$$
$$cz \geq b\ell + am.$$

We note that since $(r-1)^2 \geq 0$ for all real r, we have, by expanding and rearranging, that for all *positive* real numbers r,

$$r + \frac{1}{r} \geq 2$$

with equality if and only if $r = 1$, and we will use this observation below (for $r = b/c$, c/a, a/b). Indeed,

$$PA + PB + PC = x + y + z \geq \frac{c}{a}m + \frac{b}{a}n + \frac{c}{b}\ell + \frac{a}{b}n + \frac{a}{c}m + \frac{b}{c}\ell$$

$$= \left(\frac{b}{c} + \frac{c}{b}\right)\ell + \left(\frac{c}{a} + \frac{a}{c}\right)m + \left(\frac{a}{b} + \frac{b}{a}\right)n$$

$$\geq 2(\ell + m + n)$$

$$= 2(PL + PM + PN).$$

Moreover, we note that if equality holds, then $a = b = c$, that is $\triangle ABC$ is an equilateral triangle. $\square$

Exercise 4.19. For a point P inside a triangle $\triangle ABC$, let PL, PM, PN be the perpendiculars dropped from P to the sides. Show that

$$PA \cdot PB \cdot PC \geq 8PL \cdot PM \cdot PN.$$

Notes

Exercise 4.14 was communicated by Norman Biggs (LSE, UK).
Exercise 4.16 is based on [Alsina and Nelsen (2015)].

Chapter 5

Circles

We had defined the terms circle, its center and its radius in Section 1.6. In this chapter, we will study circles. We begin with some terminology. Recall that a circle $C(O, r)$ with center O and radius $r > 0$ is the set of points P in the plane such that $OP = r$. The set of points P in the plane such that $OP < r$ is called the *interior* of the circle, while the set of points P for which $OP > r$ is called the *exterior* of the circle. Circles having the same center are said to be *concentric*.

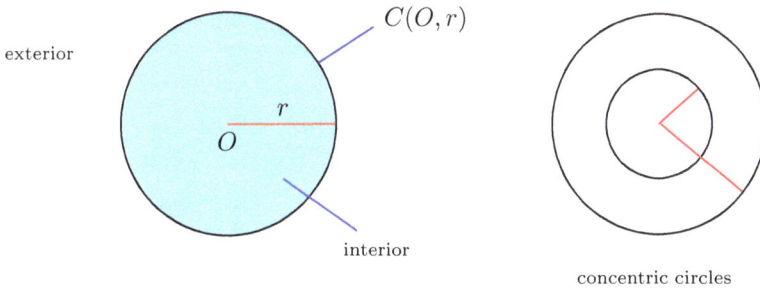

A line segment joining any two points on the circle is called a *chord*.

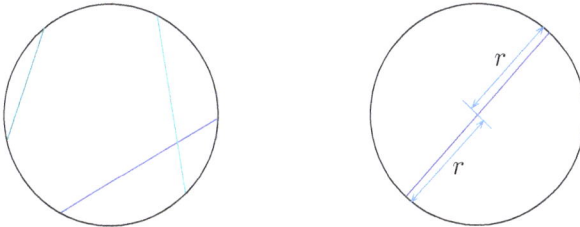

A *diameter* is a special type of chord: it is one which passes through the

center of the circle. Clearly the length of any diameter of a circle of radius
r is equal to $2r$. A diameter divides the circle into two *semicircles*.

Theorem 5.1. *The perpendicular from the center of a circle to a chord
bisects the chord.*

Proof. Let OM be the perpendicular dropped from the center O of the
circle $C(O,r)$ to its chord AB. In the right triangles $\triangle OAM$ and $\triangle OBM$,
we have $OA = OB = r$, and the side OM is common. By the RHS
Congruency Rule, $\triangle OAM \simeq \triangle OBM$, giving $AM = MB$, as wanted.   □

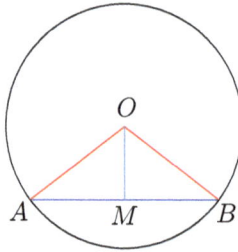

Theorem 5.2. *The line joining the center of a circle to the midpoint of a
chord is perpendicular to the chord.*

Proof. Let M be the midpoint of the chord AB of the circle $C(O,r)$.
In the two triangles $\triangle OAM$ and $\triangle OBM$, we have $OA = OB = r$, the
side OM is common, and $AM = BM$. By the SSS Congruency Rule, we
have $\triangle OAM \simeq \triangle OBM$. Hence we obtain $\angle OMA = \angle OMB$, and being
supplementary, they must each equal $90°$.   □

Corollary 5.1. *The perpendicular bisectors of two chords of a circle contain its center.*

We know that two distinct points determine a unique line passing through
them. We can now ask:

> What is the least number of distinct points needed to determine
> a unique circle passing through them?

Clearly the answer is bigger than two, since, given a pair of distinct points
A, B, any circle with its center on the perpendicular bisector of AB will
pass through A and B. See the following picture.

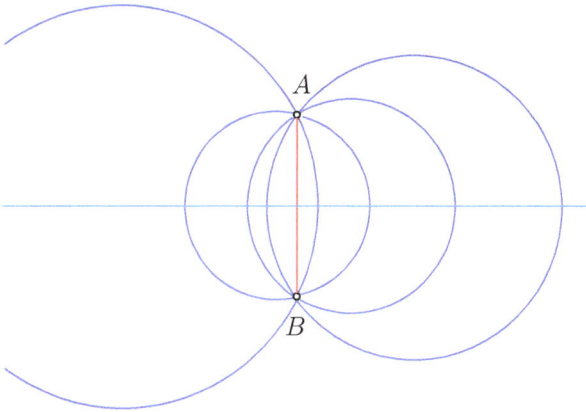

Also, three collinear points A, B, C won't do, since the center of the circle ought to lie on the perpendicular bisectors of the chords AB and BC, but these perpendicular bisectors are distinct parallel lines (and so they will never meet)! See the following picture.

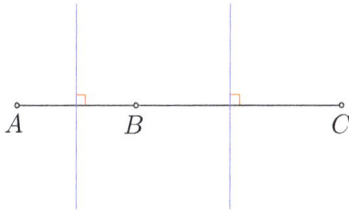

Let us now show that *three noncollinear points* uniquely determine a circle passing through them.

Theorem 5.3. *Given three noncollinear points, there is a unique circle passing through them.*

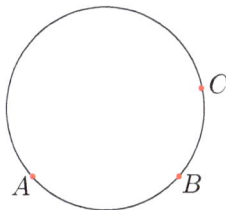

Proof. We have to show two things: that there exists a circle that passes through the three points (existence), and that there is only one such circle (uniqueness). Let us begin with the "existence" part.

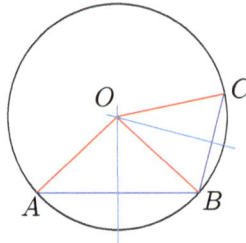

Existence: Let the given points be A, B, C. Draw the perpendicular bisectors of AB and BC. Since A, B, C are not collinear, one can see that these perpendicular bisectors are not parallel, and hence must intersect at some point O. Draw the circle $C(O, r)$ with center O and radius $r := OA$. As O lies on the perpendicular bisector of AB, $OB = OA = r$, and so B lies on $C(O, r)$. Moreover, C lies on $C(O, r)$ since O, being on the perpendicular bisector of BC, gives $OC = OB = r$. Consequently $C(O, r)$ is the circle that contains the three points A, B, C.

Uniqueness: Suppose that there is another circle $C(O', r')$ passing through A, B, C. Then the perpendicular bisectors of AB, BC will contain O' as a common point. But they contain O as well. But two nonparallel lines must intersect at only one point. Hence $O = O'$. Moreover, as A lies on $C(O, r)$ and on $C(O', r')$, it follows that $r' = O'A = OA = r$. Consequently, $C(O', r') = C(O, r)$. □

Thus the circumcircle and the circumradius of a given triangle are unique to the triangle.

Construction 5.1 (Circumcircle). *Given a triangle $\triangle ABC$, construct its circumcircle.*

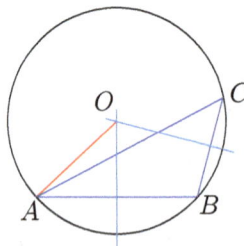

Draw the perpendicular bisectors of the sides AB, BC, and let them intersect in O. With O as center, and radius OA, draw a circle $C(O, OA)$. This is the required circumcircle. $\diamond$

From the above theorem, it follows that any two distinct circles cannot intersect each other in more than two points. Suppose on the contrary, that the two distinct circles intersect in three points A, B, C. Then these points A, B, C must necessarily be noncollinear. (If they were collinear, then the perpendicular bisectors of AB, BC must contain the centers of both circles, but these perpendicular bisectors never meet because they are distinct parallel lines.) Now, as A, B, C are noncollinear, it follows by the above result, that A, B, C pass through a unique circle, contradicting the hypothesis!

Definition 5.1 (Concyclic points). Four or more points lying on the same circle are called *concyclic*.

Definition 5.2 (Cyclic quadrilateral). A quadrilateral is said to be *cyclic* if all of its vertices lie on a circle.

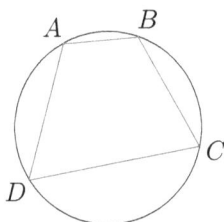

Exercise 5.1. Show that if two circles intersect in two points, then the line through their centers is the perpendicular bisector of their common chord.

Exercise 5.2. (Coming a full circle). Given an arc of a circle, provide a method for completing the circle.

5.1 Area and circumference of a circle

If we imagine a circle made of string, then we can cut the circular string at any point, stretch out the string along a line, and measure the length of the string. This length is called the *circumference* of the circle.

Theorem 5.4. *The ratio of the circumference of any circle to its diameter is a constant, denoted by* π.

Proof. We will give an argument for this fact based on the following diagram showing concentric circles, of diameters d, d', each having an inscribed polygon with n sides.

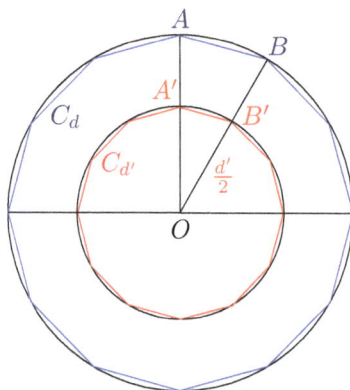

By the SAS Similarity Rule, the two triangles OAB and $OA'B'$ are similar, and thus

$$\frac{AB}{A'B'} = \frac{OB}{OB'} = \frac{d/2}{d'/2} = \frac{d}{d'}.$$

But

$$\frac{AB}{A'B'} \approx \frac{C_d/n}{C_{d'}/n} = \frac{C_d}{C_{d'}},$$

and as n becomes larger and larger, we expect the error in the above approximation to tend to 0, so that

$$\frac{C_d}{C_{d'}} = \frac{d}{d'}, \text{ that is, } \frac{C_d}{d} = \frac{C_{d'}}{d'}.$$

So the ratio of the circumference of a circle to its diameter is constant. □

Theorem 5.5. *The area of a circle of radius* r *is* πr^2.

Proof. We inscribe a regular polygon with n sides inside the circle of radius r, and triangulate it by joining the center of the circle to the vertices of the polygon. By looking at the following picture, we will justify the expression πr^2 for the area of the circle.

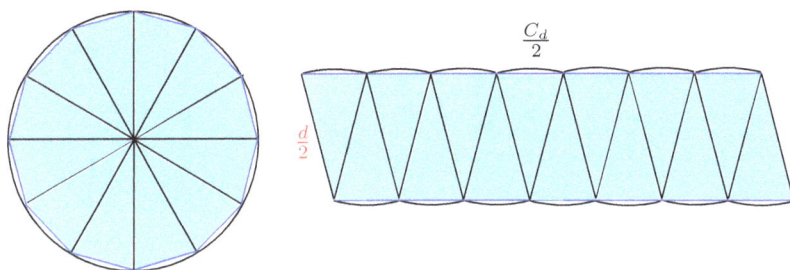

The area of the polygon is the area of the shaded parallelogram, and this is approximately the height (which differs from radius $d/2$ by a small amount) times the length of the base (which differs from half the circumference by a tiny amount). As n becomes larger and larger, we expect the errors above to go to 0, and so the area of the circle should be

$$\frac{d}{2} \cdot \frac{C_d}{2} = \frac{d}{2} \cdot \frac{\pi \cdot d}{2} = \pi \cdot \left(\frac{d}{2}\right)^2 = \pi \cdot r^2.$$

$\square$

Remark 5.1. ($\pi \notin \mathbb{Q}$). Using tools from the subject of "calculus", it can be shown that π is not a rational number. For instance, see Theorem 5.33 on pages 266-267 of [Sasane (2015)].

Exercise 5.3. (Balancing yin and yang). Given the symbol below, where smaller semicircles have diameter equal to the radius of the big circle, bisect each of the equal areas with a single straight line.

The next result gives crude estimates for π, but one usually takes its approximate value as

$$\frac{22}{7} = 3.\boxed{142857}\,\boxed{142857}\,\boxed{142857}\cdots,$$

but its precise value to 10 decimal places is $3.1415926535\cdots$.

Proposition 5.1. $3 < \pi < 2\sqrt{3} < 4$.

Proof. Consider a circle of unit radius, a regular hexagon circumscribing the circle, and a regular dodecagon[1] as shown on the left below.

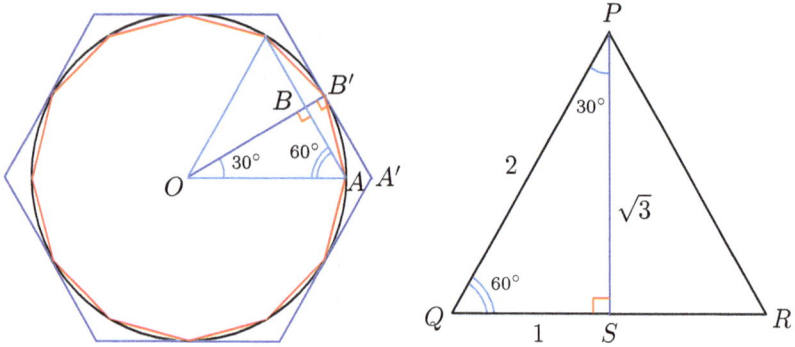

The triangles $\Delta OA'B'$, ΔOAB and ΔPQS are all similar, and so we have

$$A'B' = \frac{A'B'}{1} = \frac{A'B'}{OB'} = \frac{QS}{PS} = \frac{1}{\sqrt{3}}, \quad \text{and} \quad AB = \frac{AB}{1} = \frac{AB}{OA} = \frac{QS}{PQ} = \frac{1}{2}.$$

The interior of the circle contains the dodecagon, and so the area of the circle is bigger than the area of the dodecagon. This yields the inequality

$$\pi \cdot 1^2 > 12\Delta OAB' = 12 \cdot \frac{1}{2}OB' \cdot AB = 12 \cdot \frac{1}{2} \cdot 1 \cdot \frac{1}{2} = 3,$$

that is, $3 < \pi$. For the reverse inequality, note that since the region inside the outer hexagon covers the circle, the area of the outer hexagon is bigger than the area of the circle:

$$\pi \cdot 1^2 < 12\Delta OA'B' = 12 \cdot \frac{1}{2}A'B' \cdot OB' = 12 \cdot \frac{1}{2} \cdot \frac{1}{\sqrt{3}} \cdot 1 = 2\sqrt{3},$$

that is, $\pi < 2\sqrt{3} < 4$. Hence $3 < \pi < 4$. □

Exercise 5.4. Imagine two fixed points F, F' in the plane and a string of length L. If the two ends of the string are fixed at F, F', then the curve traced by stretching out a pencil along the string is called an *ellipse*. Thus an ellipse is the locus[2] of the point P such that $PF + PF' = $ a constant $= L$. See the following picture.

---

[1]A polygon with twelve sides.
[2]In geometry, use the word "locus" for a set of points whose location satisfies or is determined by one or more specified conditions.

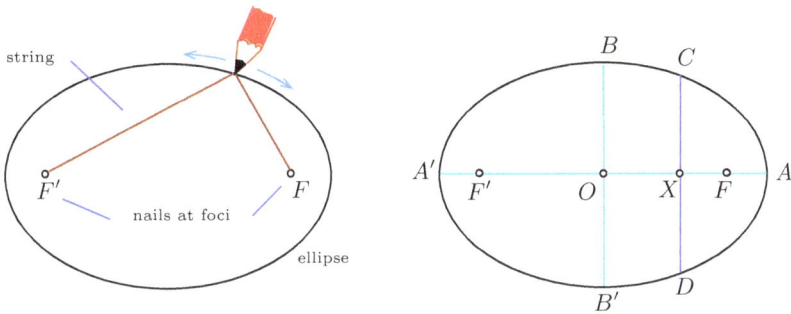

F, F' are called the *foci*[3] of the ellipse, while the line segments $A'A$, $B'B$ are called the *major axis*, respectively the *minor axis* of the ellipse.

(1) Show that if $OA =: a > b := OB$, and $OX =: x$, then the length of the vertical chord CD is given by $(2b/a) \cdot \sqrt{a^2 - x^2}$.

(2) Considering a circumscribing circle to the ellipse, and the previous part, prove that the area of the region in the interior of an ellipse is πab. What happens when b approaches a?

Exercise 5.5. (An application of the Isoperimetric Inequality). Imagine a string of length L whose two ends are tied together lying on a plane. An old problem is: For what shape of the string is the enclosed area maximized? The Isoperimetric Inequality[4] states that if the enclosed area is A, then $4\pi A \leq L^2$, with equality if and only if the shape is a circle.

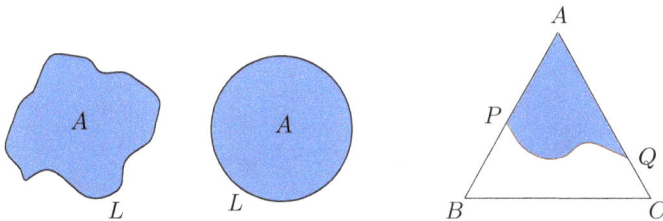

Now imagine that we have an equilateral triangle $\triangle ABC$ of side length a, and a string of length $\ell < a\sqrt{3}/2$, whose two ends lie on the sides AB and AC as shown. For what positions of the points P, Q and what shape of the curve is the area of the shaded region shown in the rightmost picture above maximized?

---

[3]Kepler (see page 66) chose this name, because he had observed that the planets in our solar system move in elliptic orbits around the sun, with the sun at one of the foci. And "focus" is Latin for "fireplace".

[4]A simple proof using calculus can be found in the article "A short path to the shortest path" by Peter Lax, *The American Mathematical Monthly*, pages 158-159, Volume 102, Number 2, February 1995.

5.2 Circular arcs

A connected portion on the circumference of the circle is called a *circular arc*.

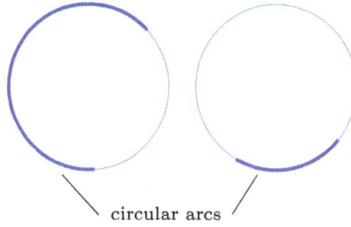

circular arcs

A chord divides the circle into two circular arcs. The smaller of these two circular arcs is called the *minor arc*, and the larger one is called the *major arc*. If the chord is a diameter, then either of the two circular arcs can be deemed to be the major or the minor arc (called semicircles).

Theorem 5.6. *In a circle, the angle subtended by any chord at the center of the circle is twice the angle subtended at any point in the major arc of the chord.*

Proof. We can have the three cases shown in the picture below. Let AB be the chord, P be the point on the major arc of the chord AB of the circle with center O. Join PO and extend it to meet the circle at the point P'.

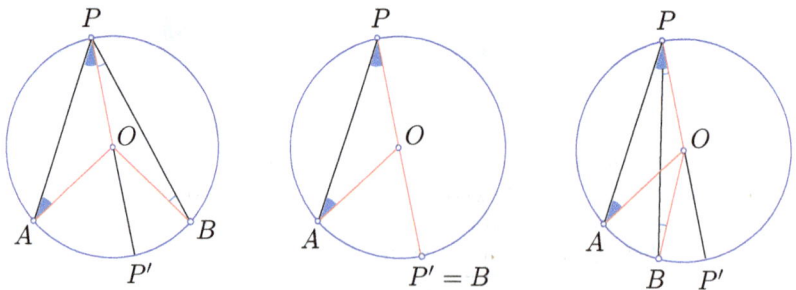

In each case, the triangle $\triangle AOP$ is isosceles, since $OA = OP$, and so we obtain $\angle OAP = \angle OPA$. Similarly, as $OP = OB$, the triangle $\triangle BOP$ is isosceles too, and $\angle OPB = \angle OBP$.

In the leftmost picture,

$$\angle AOP' = \angle OAP + \angle OPA = 2\angle OPA,$$
$$\angle BOP' = \angle OPB + \angle OBP = 2\angle OPB.$$

Adding these we obtain

$$\angle AOB = \angle AOP' + \angle BOP'$$
$$= 2\angle OPA + 2\angle OPB$$
$$= 2(\angle OPA + \angle OPB)$$
$$= 2\angle APB.$$

In the middle picture, $\angle AOB = \angle OAP + \angle OPA = 2\angle OPA = 2\angle APB$.

In the rightmost picture,

$$\angle AOP' = \angle OAP + \angle OPA = 2\angle OPA,$$
$$\angle BOP' = \angle OPB + \angle OBP = 2\angle OPB.$$

Subtracting these we obtain

$$\angle AOB = \angle AOP' - \angle BOP'$$
$$= 2\angle OPA - 2\angle OPB$$
$$= 2(\angle OPA - \angle OPB)$$
$$= 2\angle APB.$$

This completes the proof. □

Corollary 5.2. *The angle subtended by a diameter at any point of the semicircle it describes is a right angle.*

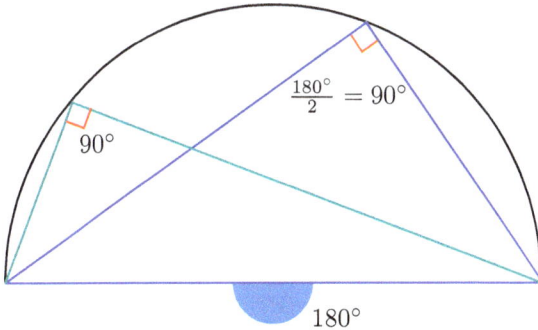

Corollary 5.3. *If a point P outside a line segment AB is such that $\angle APB = 90°$, then P lies on the circle with diameter AB.*

Proof. This is evident by considering the two pictures below where the remaining cases (P in the interior of the circle with diameter AB, or P in its exterior) are both rendered impossible.

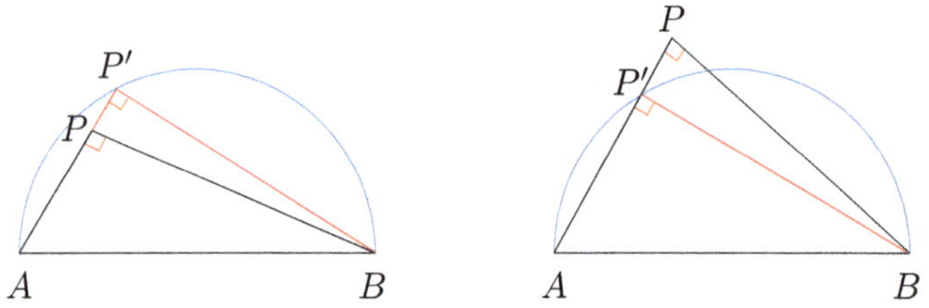

In the picture on the left, we arrive at the contradiction that

$$90° = \angle APB = \angle PP'B + \angle P'BP = 90° + \angle P'BP > 90°,$$

while in the picture on the right, we arrive at the contradiction that

$$90° = \angle AP'B = \angle P'PB + \angle PBP' = 90° + \angle PBP' > 90°.$$

This completes the proof. $\qquad\square$

Exercise 5.6. (Apollonius's Circle). ($*$) Let A, B be two fixed points in the plane, and let $r > 1$. It is required to find the locus of a mobile point P such that $PA : PB = r$. In order to find this locus, proceed as follows.

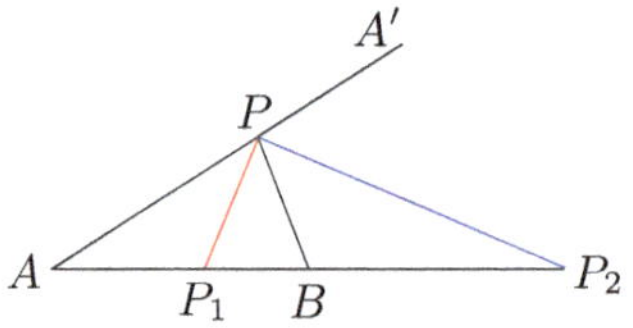

Let P be a point not on the line containing AB. Draw the angle bisector of $\angle APB$ and suppose it meets AB in P_1. Extend AP to a point A'. Draw the angle bisector of $\angle A'PB$ and suppose that it meets AB extended at P_2. Answer the following questions.

(1) Does P_1 lie on the locus traced by P?

(2) Does P_2 lie on the locus traced by P?

(3) What is $\angle P_1PP_2$?

(4) What is the locus of P?

(5) What if $r < 1$? What if $r = 1$?

Exercise 5.7. (What does means mean? Means means means!) Suppose that line segments with lengths a, b where $a > b > 0$ are given. Draw a line segment AB of length a, and extend it to a point A' so that the length of BA' is b. Construct the midpoint O of AA'. Draw a semicircle with center O and radius OA'. Erect a perpendicular to AB at B, which meets the semicircle at B'. Erect a perpendicular to AB at O which meets the semicircle at O'. Join $O'B$ and OB'. Drop a perpendicular from B onto OB', meeting it in the point C.

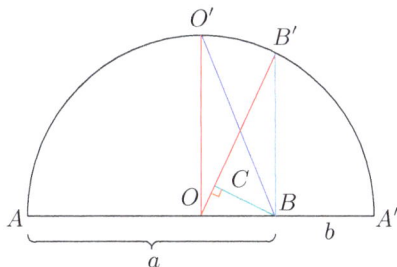

Show the following claims:

(1) $OO' = \dfrac{a+b}{2}$, the arithmetic mean AM_{ab} of a and b.

(2) $BB' = \sqrt{ab}$, the geometric mean GM_{ab} of a and b.

(3) $O'B = \sqrt{\dfrac{a^2+b^2}{2}}$, the quadratic mean QM_{ab} of a and b.

(4) $B'C = \dfrac{2}{1/a + 1/b}$, the harmonic mean HM_{ab} of a and b.

(5) Show that $\mathrm{QM}_{ab} \geq \mathrm{AM}_{ab} \geq \mathrm{GM}_{ab} \geq \mathrm{HM}_{ab}$.

Exercise 5.8. The paragraph at the beginning of the previous exercise gives a method for constructing the various means of two given lengths. For example using the method for constructing the geometric mean, given a rectangle, we can construct a square with area equal to the rectangle[5]. Taking one of the numbers as 1, we can construct square roots. So given a unit length, one can construct for example the number $\sqrt{(\sqrt{28} + \sqrt{3})/(\sqrt{1976} + \sqrt[32]{399})}$.

[5]On the other hand, the ancient geometers considered the problem of "Squaring the circle", where given a circle, it is required to construct a square of the same area as that enclosed by the circle. A real number α is called *algebraic* if there exists a nonzero polynomial p with integer coefficients such that $p(\alpha) = 0$. For example, all rational numbers are algebraic, and so are some irrational numbers like $\sqrt{2}$. It was known that squaring the circle would be an impossible Euclidean construction with a straight edge and a compass if π is nonalgebraic (also known as transcendental numbers), and in 1882, this was shown by Lindemann as a consequence of his more general result saying that e^α is transcendental for every nonzero algebraic number α. As $e^{i\pi} = -1$, which is algebraic, it follows that π ought to be nonalgebraic.

We will meet yet another application of the constructibility of the geometric mean when we discuss drawing a tangent to a circle from a point outside the circle when the center of the circle is not known, in Construction 5.4.

Besides the method for constructing the harmonic mean given in the previous exercise, here is another way to construct the harmonic mean. Given line segments of lengths a, b where $a \geq b > 0$, draw any trapezium $ABCD$ with the parallel sides AB, CD having lengths a, b, respectively. (This amounts to drawing AB first, then taking any point C outside the line containing AB, and drawing a line through C parallel to AB. The point D can be cut on this line through C such that $CD = b$.) Join A to C, and also B to D, to obtain the diagonals AC and BD, and suppose that the diagonals meet at P. Draw a line through P parallel to the sides AB, CD, and let them meet AD, BC in Q and R. Then show that the length of QR is the harmonic mean of a and b.

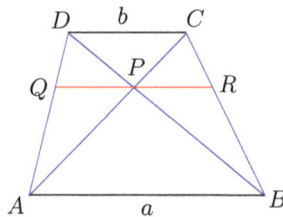

Exercise 5.9. Let B, C be two fixed points in the plane and let A be a variable point. Find the locus of the incenter of $\triangle ABC$.

Exercise 5.10. (Geometric diagram for relativistic velocity addition). Imagine a train moving at speed u with respect to the ground (as reckoned by someone sitting on the ground), and further that a person P is running with a speed v on the train (as reckoned by somebody sitting in the train). Before 1905, Newtonian physics dictated that the speed of the person P as observed by someone on the ground is $u + v$, while we now know better; the relativistic formula for velocity addition says that the speed should be $(u \oplus v) := (u + v)/(1 + uv)$, in units in which the speed of light is 1.

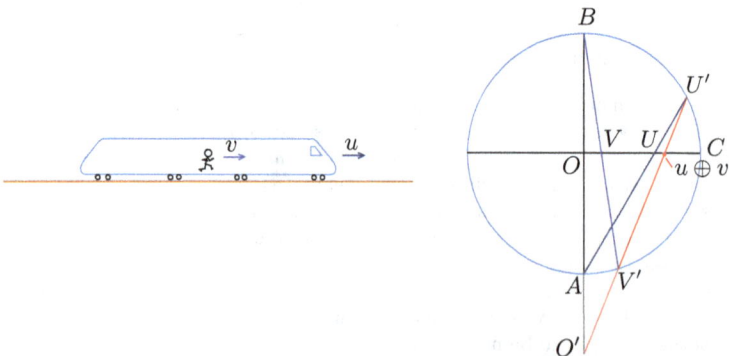

The picture on the right above gives a diagram for the construction of $u \oplus v$. Draw a circle with center O and radius 1. Mark points U, V at distances u, v from O along the radius OC perpendicular to a diameter AB. Let the line joining B to V meet the circle at V', and let the line joining A to U meet the circle at U'. Then $u \oplus v$ is the distance from O of the point of intersection of $U'V'$ with the radius OC. We will justify this in this exercise. Determine AU, BV', UU', VV' in terms of u, v. Let the extension of $U'V'$ meet the extension of AB at O'. Apply Menelaus's Theorem in $\triangle BOV$ and $\triangle AOU$ with the line $U'V'$ to determine OW, where W is the point of intersection of $U'V'$ with OC.

The geometric diagram can be used to give visual proofs of identities $u \oplus 0 = u$, $u \oplus v = v \oplus u$, $u \oplus 1 = 1$. Convince yourself of these. Based on the geometric diagram, justify that $u \oplus v \approx u + v$ if $u, v \ll 1$ (that is, when u, v are much smaller than 1).

Corollary 5.4. *The angle subtended by any chord of a circle at any point on the major arc it describes is the same.*

Corollary 5.5. *The angle subtended by a chord at any point of the minor arc is supplementary to the angle subtended at any point of the major arc.*

Proof. Let AB be the chord of a circle with center O, and let P be any point on the minor arc described by AB. Join OA, OB and OP.

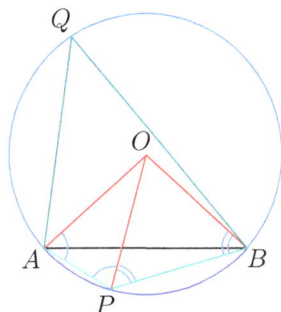

In the isosceles triangles $\triangle OAP$ and $\triangle OPB$,

$$\angle APO = 90° - \frac{\angle AOP}{2}, \text{ and } \angle OPB = 90° - \frac{\angle BOP}{2}.$$

Hence if Q is any point on the major arc, we have $\angle AOB = 2\angle AQB$, and

$$\angle APB = \angle APO + \angle OPB = 180° - \frac{\angle AOP + \angle BOP}{2}$$

$$= 180° - \frac{\angle AOB}{2} = 180° - \angle AQB.$$

$\square$

Corollary 5.6. *The angle subtended by any chord of a circle at any point on the minor/major arc it describes is the same.*

Proof. In the above, first keep the point P on the minor arc of the chord fixed and let Q be mobile over the major arc. Then we see from Corollary 5.5 that the angle subtended by a chord at any point on its major arc is the same. Similarly, by keeping Q fixed and letting P be mobile over the minor arc, we obtain the constancy of angle subtended by a chord of a circle at any point on the minor arc too. □

As a converse to Corollary 5.2, we have the following result; see also Corollary 5.3.

Theorem 5.7. *A chord of a circle which subtends a right angle at a point of its major (or its minor) arc is a diameter.*

Proof. Let the chord be AB of the circle $C(O,r)$, and suppose that it subtends a right angle at the point P of its major/minor arc. Then A, B, O must be collinear since we have $\angle AOB = 2 \cdot 90° = 180°$. □

Exercise 5.11. (∗) Given two points A, B inside a circle C, show that the point on the circle C where AB subtends the greatest angle is the point of contact of the smaller of the two circles through A and B which touch the given circle C internally.

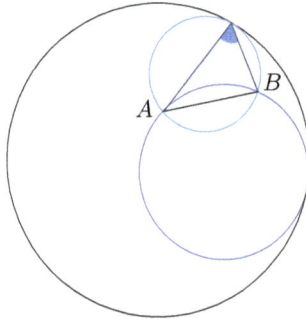

Theorem 5.8 (9-point circle). *In any triangle, the following 9 points:*
- ··· *midpoints of the sides of the triangle,*
- ··· *the feet of the altitudes, and*
- ··· *the midpoints of the lines joining the orthocenter to the vertices,*

are concyclic[6]. Moreover, the radius of the 9-point circle is half the circumradius of the triangle.

---

[6]The circle passing through these 9 points is called the 9-point circle *of the triangle.*

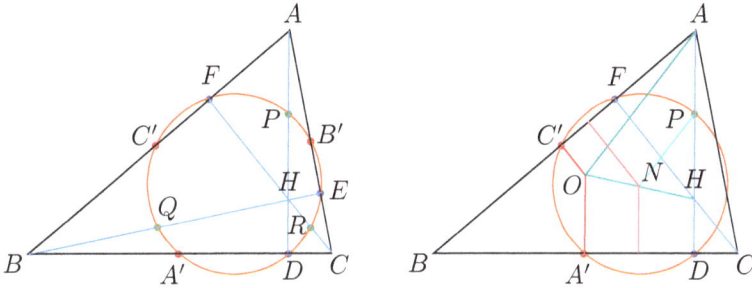

Proof. Let the triangle be $\triangle ABC$. We will use the following notation.

$$A', B', C' \; : \; \text{midpoints of } BC, CA, AB,$$
$$D, E, F \; : \; \text{feet of the altitudes of } \triangle ABC \text{ dropped from } A, B, C,$$
$$H \; : \; \text{orthocenter of } \triangle ABC,$$
$$P, Q, R \; : \; \text{midpoints of } AH, BH, CH.$$

By the Midpoint Theorem, $C'P \parallel BH$ and $C'A' \parallel AC$. But as $BH \perp AC$, it follows that $C'A' \perp C'P$, that is, $\angle A'C'P = 90°$. Similarly we have $\angle A'B'P = 90°$ too. Also, we know that $\angle PDA' = 90°$. So by Theorem 5.7, if C_* is the circle with diameter $A'P$, then the points C', B', D lie on C_*. In other words, the circle C_* through A', B', C' passes through P and D. Similarly, C_* passes through Q, E and through F, R as well. (And $B'Q$ and $C'R$ are diameters of C_* too.)

Let O be the circumcenter of $\triangle ABC$. Then OA is its circumradius. The perpendicular bisector of $C'F$ meets the side OH of the trapezium $C'OHF$ in the midpoint N of OH. Similarly, the perpendicular bisector of $A'D$ meets OH in the midpoint N too. So the perpendicular bisectors of the chords $C'F$ and $A'D$ of C_* meet at N, implying that N is the center of C_*, and so the radius of C_* is NP. But in $\triangle AOH$, $OA = 2 \cdot NP$, where the last equality follows from the Midpoint Theorem. Hence the radius of C_* is $NP = OA/2$, half the circumradius of $\triangle ABC$. $\square$

Remark 5.2. Note that in the proof above, we showed that the midpoint N of OH is the center of the 9-point circle C_*. But O, H determine the Euler line (see Theorem 5.11 below), which also contains the centroid G. Hence the *four* points, G, H, O, N are collinear, that is, for any triangle, the centroid, the orthocenter, the circumcenter, and the center of the 9-point circle, are collinear, and lie on the Euler line of the triangle.

Theorem 5.9. *If a line segment joining two points subtends equal angles at two other points lying on the same side of the line containing the line segment, then the four points are concyclic.*

Proof. Let the given line segment be AB, and let the other two given points (where AB subtends equal angles) be labelled as C, D. Draw the circle passing through A, B, C. If D doesn't lie on this circle, then consider the point $D' \neq D$ of intersection of the circle with AD (possibly extended).

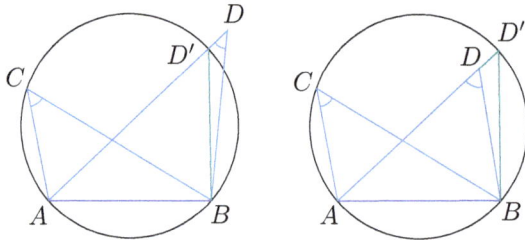

Then $\angle ACB = \angle AD'B$. But $\angle ACB = \angle ADB$ (given). So we have $\angle AD'B = \angle ADB$. On the other hand, by looking at the triangle $\triangle BDD'$,

$$\text{either } \angle AD'B > \angle ADB \quad \text{(left picture above)}$$
$$\text{or } \angle AD'B < \angle ADB \quad \text{(right picture above)}.$$

In either case we have a contradiction, and so it must be the case that D lies on the circle passing through A, B, C, that is, the points A, B, C, D are concyclic.   □

Exercise 5.12. If points P, Q on either side of the line containing a line segment AB are such that $\angle APB + \angle AQB = 180°$, then A, P, B, Q are concyclic.

Theorem 5.10 (Simson's Line). [7] *Let P be any point on the circumcircle of a triangle $\triangle ABC$, and let L, M, N be the feet of the perpendiculars dropped from any point P to the three sides BC, CA, AB. Then the points L, M, N are collinear.*

(The line containing L, M, N is called the *Simson's line* or *pedal line* of the triangle $\triangle ABC$ with respect to the point P.)

---

[7]After Robert Simson (1687-1768), a Scottish professor of mathematics at the University of Glasgow.

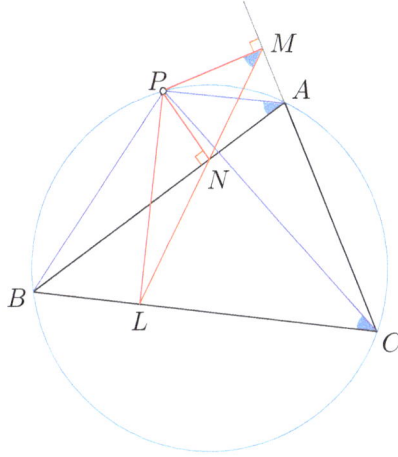

Proof. From the point P on the circumcircle of $\triangle ABC$, drop perpendiculars to the sides CA and AB, meeting them in M, N, respectively. Join M to N and extend it to meet BC in L. We need to show that PL is perpendicular to BC.

We first note that in the quadrilateral $PMAN$, since the opposite angles $\angle PMA$ and $\angle PNA$ are both $90°$, $PMAN$ is cyclic. Thus we conclude that $\angle PML = \angle PAB$. But the chord BP on the circumcircle of $\triangle ABC$ subtends the same angle at A and C, so that $\angle PAB = \angle PCL$. From the above, we conclude that $\angle PML = \angle PCL$. So the segment PL subtends the same angles at M and C. Hence the points P, M, C, L are concyclic. Since $\angle PMC = 90°$, we have $\angle PLC = 180° - 90° = 90°$ too. This completes the proof. $\qquad\square$

Theorem 5.11 (Euler line). [8] *In a nonequilateral triangle, the circumcenter O, the centroid G, and the orthocenter H are collinear, and moreover, $GH = 2 \cdot OG$.*

(The line determined by O, G, H is called the *Euler line* of the nonequilateral triangle. In the case of an equilateral triangle, the three points O, G, H coincide.)

Proof. Let A' be the midpoint of the side BC, and let AD, CF be altitudes dropped from A and C, respectively, to the sides BC, AB. Let AP be a circumdiameter.

---

[8] After Leonhard Euler (pronounced "Oiler"; 1707-83), a Swiss mathematician.

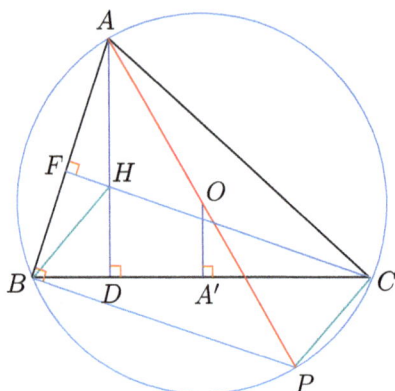

Since $\angle ABP = 90° = \angle AFC$, it follows that $FC \parallel BP$. Similarly, the extension of BH is perpendicular to AC, and PC is also perpendicular to AC, implying that $BH \parallel PC$. Thus $BHCP$ is a parallelogram, and so its diagonals BC and HP bisect each other, and this bisection happens at A' since A' is the midpoint of BC. So A' is the midpoint of HP too. Now in $\triangle AHP$, the line OA' passing through the midpoint A' of one side is parallel to the side AH, and so by the Midpoint Theorem 2, $2 \cdot OA' = AH$.

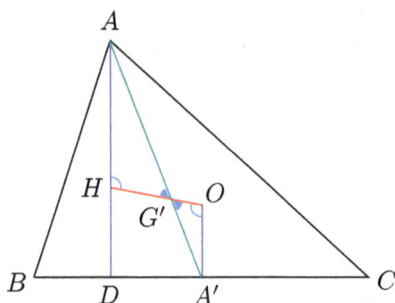

Join AA', and suppose that it meets OH at G'. Then the opposite angles $\angle AG'H$ and $\angle A'G'O$ are equal, and also, since $AH \parallel OA'$, the interior alternate angles $\angle AHG'$ and $\angle G'OA'$ are equal. By the AA Similarity Rule, $\triangle AHG' \sim \triangle A'OG'$. As $2 \cdot OA' = AH$, it follows that $AG' = 2 \cdot G'A'$ and $HG' = 2 \cdot G'O$. Since G' divides the median AA' in the ratio $2 : 1$, G' must be the centroid G, and $HG = HG' = 2 \cdot G'O = 2 \cdot GO$. $\qquad\square$

Fermat's point

Theorem 5.12. *Consider a triangle $\triangle ABC$ where no angle is bigger than $120°$. Describe equilateral triangles $\triangle ABC'$, $\triangle BCA'$ and $\triangle CAB'$ on AB, BC, CA. Then:*
  (1) *AA', BB', CC' are concurrent.*
  (2) *$\angle AFB = \angle BFC = \angle CFA = 120°$.*
  (3) *For all points P in the interior of $\triangle ABC$,*
      $$PA + PB + PC \geq FA + FB + FC.$$

The point of concurrency is called the *Fermat*[9] *point* of the triangle.

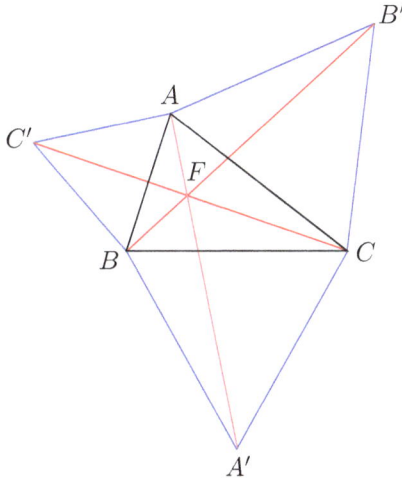

Proof.
(1): In order to prove the concurrency, join BB' and CC', and let them meet in F. Join F to A and also to A'. We need to show that A, F, A' are collinear.

First, we observe that $\triangle AC'C \simeq \triangle ABB'$ by the SAS Congruency Rule (since we have $AC' = AB$, $AC = AB'$, and the included angle is $\angle A + 60°$).

By the congruency of $\triangle AC'C$ and $\triangle ABB'$, $\angle AC'F = \angle ABF$. So the points A, C', B, F are concyclic. Hence $\angle C'FB = \angle C'AB = 60°$ and $\angle C'FA = \angle C'BA = 60°$. Consequently, $\angle AFB = 120°$.

Similarly, again by the congruency of $\triangle AC'C$ and $\triangle ABB'$, we also conclude that $\angle AB'F = \angle ACF$. So the points A, B', C, F are concyclic.

---
[9] After the 17th century French mathematician Pierre de Fermat.

Hence we have $\angle B'FC = \angle B'AC = 60°$ and $\angle AFB' = \angle ACB' = 60°$. So $\angle AFC = 120°$ too.

Thus $\angle BFC = 360° - 2 \cdot 120° = 120°$. Since $\angle BFC = 120°$ and $\angle BA'C = 60°$ are supplementary, we conclude that the points A', B, F, C are concyclic. So we have that $\angle BFA' = \angle BCA' = 60°$. Finally, we obtain $\angle AFB + \angle A'FB = 120° + 60° = 180°$, and so A, F, A' are collinear.

(2): Note that in (1) we've also shown $\angle AFB = \angle BFC = \angle CFA = 120°$.

(3): Draw three lines ℓ_A, ℓ_B, ℓ_C which are respectively perpendicular to FA, FB, FC, and which pass through the points A, B, C, respectively. Suppose that their pairwise meeting points are C'', A'', B'' as shown. Using the perpendicularity and the fact (2) established above, it follows that $\angle A''$, $\angle B''$ and $\angle C''$ are each equal to $60°$, and so $\triangle A''B''C''$ is equilateral.

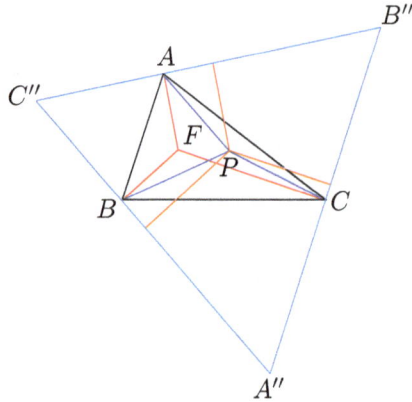

Now suppose that P is any point inside $\triangle ABC$. By Viviani's Theorem[10], it follows that the sum of the lengths of the perpendiculars dropped from P onto the three sides of $\triangle A''B''C''$ equals $FA + FB + FC$ on the one hand, and (by Pythagoras's Theorem) surely less than $PA + PB + PC$. This completes the proof. □

In the above result, we made the assumption that all the triangles in $\triangle ABC$ are less than $120°$. It is natural to ask what happens when one of the angles, say $\angle A$ is bigger than or equal to $120°$. Well, in that case, AA', BB', CC' meet outside the triangle, and the solution to the optimization problem in item (3), namely the point P inside $\triangle ABC$ minimizing $PA + PB + PC$ is A; see the following paragraph.

---

[10]See Exercise 3.21.

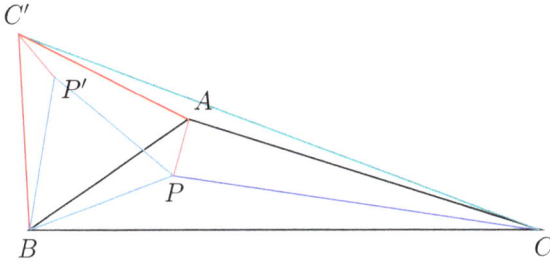

In order to show the previous claim, take any point P inside $\triangle ABC$, and rotate the triangle $\triangle APB$ by $60°$ about the point B, as shown, so that A is sent to the point C', and P is sent to P'. Then

$$PA + PB + PC = C'P' + P'P + PC > AC' + AC = AB + AC,$$

where the inequality follows from the Dogleg Rule (Lemma 2.1).

Construction of a regular pentagon; golden ratio

Suppose, given the side length AB, we wish to construct a regular pentagon $ABCDE$.

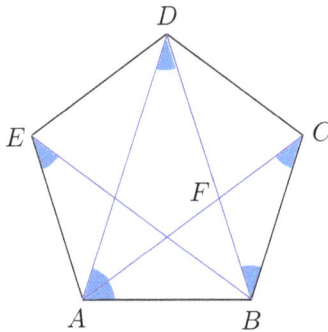

In the triangles $\triangle ABC$ and $\triangle BAE$, $AE = BC$, side AB is common, and $\angle EAB = \angle ABC$. Thus $\triangle ABC \simeq \triangle BAE$ by the SAS Congruency Rule, and so $\angle AEB = \angle ACB$. If we look at the circle passing through A, B, C, then the chord AB subtends the same angle at the points E and C. Thus E must lie on this circle, and the points A, B, C, E are concyclic. Similarly, A, B, C, D are also concyclic. Hence A, B, C, D, E are concyclic. In particular, $\angle ADB = \angle ACB =: \alpha$.

Now we identify the following isosceles triangles by the symmetry of the regular pentagon. First, let F be the point of intersection of AC and BD.

The triangles $\triangle ADB$ (with $AD = BD$), $\triangle ABC$ (with $AB = BC$), and $\triangle BFC$ (with $BF = CF$) are all isosceles triangles. So, from the above, $\alpha = \angle ADB = \angle ACB = \angle CAB = \angle FBC$. The chord DC subtends the same angles at A and B, and so we have $\angle DAC = \angle FBC = \alpha$. Finally, $\angle AFB = \angle FCB + \angle FBC = 2\alpha$.

Now in $\triangle ADB$ and $\triangle FAB$, we have that $\angle ADB = \angle FAB = \alpha$, and also $\angle DAB = \angle AFB = 2\alpha$. By the AA Similarity Rule, $\triangle ADB \sim \triangle FAB$. With $x := AD = BD = AC$, $FB = FC = AC - AB = AD - AB = x - AB$, and

$$\frac{x}{AB} = \frac{AD}{AB} = \frac{AF}{FB} = \frac{AB}{FB} = \frac{AB}{x - AB},$$

that is, $x^2 - ABx - AB^2 = 0$. Discarding the negative root, we obtain

$$AD = BD = x = \frac{1 + \sqrt{5}}{2} \cdot AB.$$

By constructing the right angled triangle with nonhypotenuse side lengths AB and $2AB$, we obtain the length $\sqrt{5}AB$ as the length of the hypotenuse. Hence we can also construct (by adding the length AB to the length $\sqrt{5}AB$, and bisecting the resulting line segment) the length x.

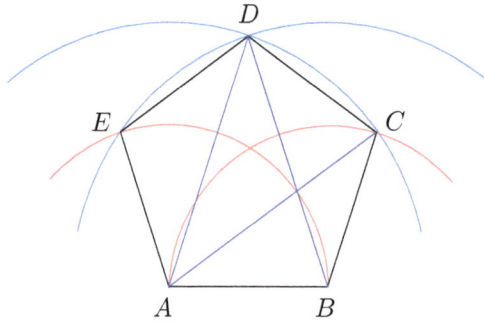

Now the construction of the pentagon $ABCDE$ is straightforward:

(1) With A as center and radius x, draw a circle $C(A, x)$. Similarly, draw the circle $C(B, x)$. Let D be one of the intersection points of $C(A, x)$ and $C(B, x)$.
(2) On the same side of AB as the point D, let C be the intersection point of the circle $C(A, x)$ and $C(B, AB)$, and let E be the intersection point of the circle $C(B, x)$ and $C(A, AB)$.

Note that since each internal angle of a regular pentagon is $108°$, besides this angle, we can also construct the angles $180° - 108° = 72°$, $72/2 = 36°$, $18°$, $9°$, $72 + 9 = 81°$ and so on. The number

$$\tau := \frac{1 + \sqrt{5}}{2} \approx 1.618 \cdots$$

is called the *golden ratio*, sometimes also called the *divine proportion*. This ratio is believed to create geometrical figures of particularly pleasing proportions, for example, the *golden rectangle*, with sides in the ratio $1 : \tau$. When, using the smaller side, a square is separated from the golden rectangle, we obtain yet another golden rectangle, thanks to the relation $\tau^2 - \tau - 1 = 0$, that is, $\tau - 1 = 1/\tau$.

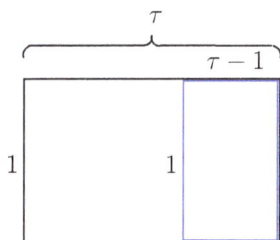

Some artists[11] have used the golden ratio in their art. The golden ratio also appears naturally in nature, and in mathematics, for example as the "limiting value" of the ratio of successive Fibonacci numbers (see Exercise 5.15 below): that is, for large n, $F_{n+1} : F_n \approx \tau$.

Exercise 5.13. If the sides of a right triangle are in geometric progression $1 : r : r^2$ with $r > 1$, then determine the value of r.

Exercise 5.14. Give a geometric proof that τ can't be rational. Conclude that $\sqrt{5}$ is not rational either.

Exercise 5.15. (Fibonacci[12] numbers). The *Fibonacci sequence* of numbers is generated by the recurrence rule $F_0 = F_1 = 1$, and for $n > 1$, $F_{n+1} = F_n + F_{n-1}$. Thus the first few terms of the sequence are $1, 1, 2, 3, 5, 8, 13, 21, 34, 55, 89, 144, \cdots$.

(1) Assuming that there is real number L such that $F_{n+1} : F_n \approx L$ for large n, use the recurrence relation to show that L must be τ.

(2) A French mathematician travelling by car in the UK notices the following curiosity: in order to approximately convert a Fibonacci number of miles into kilometers, all one needs to do is to take the next Fibonacci number! Can you explain why?

---

[11] For example, Salvador Dali.
[12] After the Italian mathematician Fibonacci (c.1170-1250).

(3) Consider the Fibonacci spiral made by putting together squares of side F_n, one at a time, in a spiral manner, getting a $F_n \times F_{n+1}$ rectangle at each stage, as shown below. Show, by comparing areas, that for all n, we have $F_0^2 + F_1^2 + \cdots + F_n^2 = F_n F_{n+1}$.

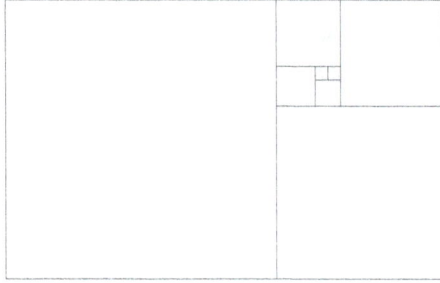

Theorem 5.13 (Ptolemy). [13] *A quadrilateral $ABCD$ is cyclic if and only if the product of the diagonals is the sum of the products of the opposite pairs of sides: $AC \cdot BD = AB \cdot CD + BC \cdot AD$.*

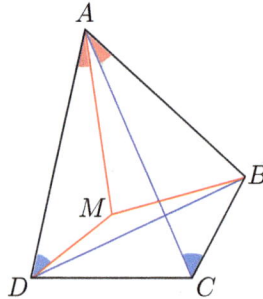

Proof. Inside a quadrilateral $ABCD$, let the point M be such that

$$\angle ADM = \angle ACB \text{ and } \angle DAM = \angle CAB.$$

Then $\triangle ADM \sim \triangle ACB$, and so $AD/AC = MD/BC$, that is,

$$AD \cdot BC = AC \cdot MD. \tag{5.1}$$

Also, $AM/AB = AD/AC$, or

$$\frac{AM}{AD} = \frac{AB}{AC}, \tag{5.2}$$

[13]Ptolemy (AD 90 to 168) was a Greco-Egyptian mathematician of Alexandria.

and

$$\angle DAC = \angle DAM + \angle MAC = \angle CAB + \angle MAC = \angle MAB. \qquad (5.3)$$

From (5.2) and (5.3), $\triangle DAC \sim \triangle MAB$ (SAS Similarity Rule). So we conclude that $AC/AB = CD/BM$, that is,

$$AB \cdot CD = AC \cdot BM. \qquad (5.4)$$

From (5.1) and (5.4), we obtain

$$AD \cdot BC + AB \cdot CD = AC \cdot (BM + MD) \geq AC \cdot BD. \qquad (5.5)$$

In the above, we have used $BM + MD \geq BD$, and we note that we have equality here if and only if M lies on BD.

"Only if" part: If $ABCD$ is cyclic, then AB subtends equal angles at C and D, and so $\angle ACB = \angle ADB$. But $\angle ACB = \angle ADM$ by construction. Thus $\angle ADB = \angle ADM$, and this implies that M lies on BD. So (5.5) gives $AC \cdot BD = AB \cdot CD + BC \cdot AD$, as required.

"If" part: If $AC \cdot BD = AB \cdot CD + BC \cdot AD$, then the equality in (5.5) gives us that $BM + MD = BD$, and so M lies on BD. But then we have $\angle ADB = \angle ADM = \angle ACB$. By looking at the circle passing through the points A, B, C, it follows that D lies on this circle too, thanks to the fact that $\angle ADB = \angle ACB$. $\qquad \square$

Exercise 5.16. Deduce Pythagoras Theorem using Ptolemy's Theorem.

Exercise 5.17. Show that for any point P on the circumcircle of an equilateral triangle $\triangle ABC$, the sum of the smaller two among PA, PB, PC is equal to the third one.

Theorem 5.14. *Let P be any point inside a given circle $C(O, r)$, and let ℓ be any line passing through P, intersecting the circle at the points A_ℓ, B_ℓ. Then the product $A_\ell P \cdot PB_\ell$ is a constant (depending only on $C(O, r)$ and P).*

Later on, in Theorem 5.18, we will also learn an analogous result when the point P lies *outside* the circle $C(O, r)$.

Proof. Look at the following picture, where the chords $AB, A'B'$ intersect at P. We would like to show that $AP \cdot PB = A'P \cdot PB'$. Join A to A', and B to B'.

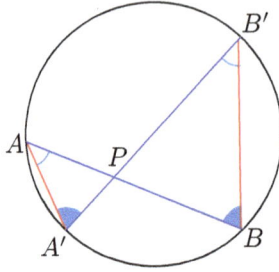

We claim that $\triangle APA' \sim \triangle B'PB$. Indeed, the chord $A'B$ subtends equal angles at A, B', and so we have that $\angle A'AP = \angle BB'P$. Similarly by considering the chord AB', we also conclude that $\angle AA'P = \angle B'BP$. Hence by the AA Similarity Rule, we obtain that $\triangle APA' \sim \triangle B'PB$. Consequently,

$$\frac{AP}{PB'} = \frac{A'P}{PB},$$

that is, $AP \cdot PB = A'P \cdot PB'$. $\square$

Later on, in Theorem 5.18, we will learn a similar result when P lies outside the circle.

Exercise 5.18. (A simple construction of τ). If an equilateral triangle $\triangle ABC$ is inscribed in a circle and the line segment DE joining the midpoints D, E of the two sides AB, AC is extended to intersect the circle at a point F, then $DE : EF = \tau$, the golden ratio.

Construction 5.2 (Triangle from altitude lengths). *Given lengths of the three altitudes of a triangle, construct the triangle.*

As the area of a triangle is half the base length times the corresponding altitude, we have the relation

$$a \cdot h_a = b \cdot h_b = c \cdot h_c,$$

for the sides a, b, c and the altitudes h_a, h_b, h_c. On the other hand, if we think of three chords $AB, A'B', A''B''$ intersecting at a point P in a circle such that $AP = h_a$, $A'P = h_b$ and $A''P = h_c$, then it follows that PB, PB', PB'' are proportional to a, b, c. So at least we can construct a triangle similar to our required triangle.

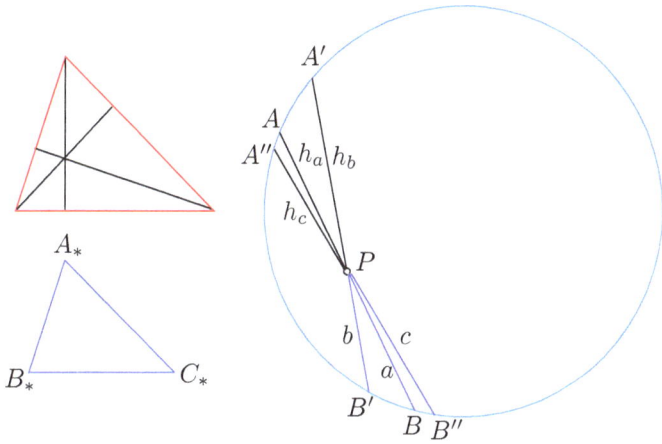

The procedure is this: Draw a large enough circle, and take a point P close enough to the boundary. With P as center and radius h_a, h_b, h_c respectively, draw circular arcs intersecting our big circle at the points A, A', A'', respectively. Extend $AP, A'P, A''P$ to meet the circle again at the points B, B', B'', respectively. Draw a triangle with the side lengths PB, PB', PB'', and call it $\triangle A_* B_* C_*$.

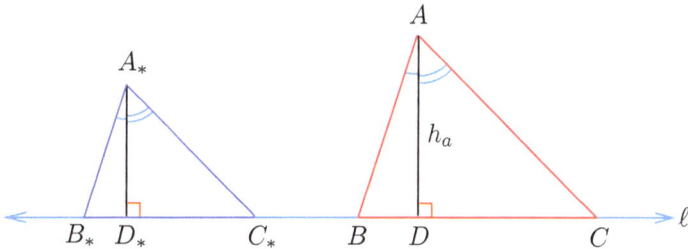

Now the task is to construct a similar triangle $\triangle ABC$ to $\triangle A_* B_* C_*$ such that the altitudes in $\triangle ABC$ have the given lengths h_a, h_b, h_c. (Without loss of generality, we continue with our construction assuming that A_* is the vertex that has the largest angle in $\triangle A_* B_* C_*$—then the altitude dropped from A_* onto $B_* C_*$ will meet it somewhere *inside* $B_* C_*$.) Drop the altitude $A_* D_*$ from A_* to $B_* C_*$. On the line ℓ containing $B_* C_*$, erect a perpendicular AD of height h_a. Mark the point B on ℓ such that we have $\angle BAD = \angle B_* A_* D_*$. Similarly on the other side of D, mark the point C such that $\angle CAD = \angle C_* A_* D_*$. Then $\triangle ABC$ is the required triangle. $\diamond$

Remark 5.3. Impossibility of the construction of a triangle given the angle bisector lengths. Earlier we had seen that given the median lengths, we can construct the triangle, and we have seen above that the triangle construction is also possible when the altitude lengths are given. One may then ask the natural question:

Can one construct a triangle given its angle bisector lengths?

The problem has a long history, and was proposed by Brocard in *Mouvelle Correspondance Mathématique* in 1875. The challenge met with no solution in the journal's lifetime (1875-80), and a proof[14] of the impossibility of the construction in general was given by Korselt in 1897. There were also two PhD theses written on this problem.

Theorem 5.15 (Euler). *In any triangle, if O, I are the circumcenter and incenter, respectively, and R, r are the circumradius and inradius, respectively, then $OI^2 = R^2 - 2Rr$.*

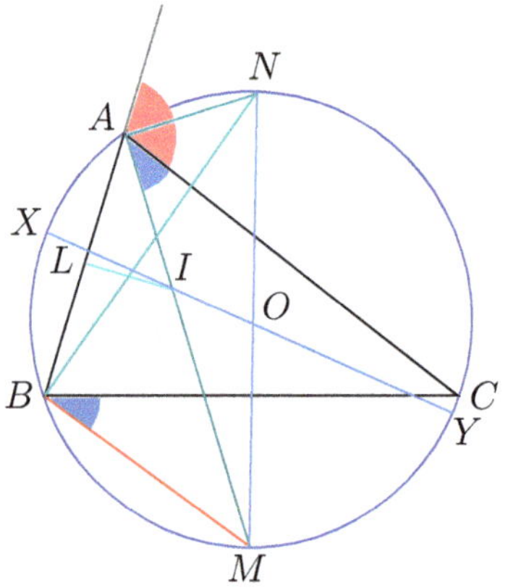

Proof. Let the triangle have vertices A, B, C, and let AI extended meet the circumcircle at the point M. Then we have $\angle MBC = \angle MAC = \angle A/2$. So $\angle MBI = \angle A/2 + \angle B/2 = \angle BAI + \angle ABI = \angle BIM$. Hence $\triangle IBM$ is isosceles, and $BM = IM$.

Let the external angle bisector of $\angle A$ meet the circumcircle at N. Then we have $\angle MAN = 90°$, and so MN is a diameter. Now if IO extended

---

[14]A proof can also be found as the solution to problem 454 in *Crux Mathematicorum*, Vol. 6, No. 4, April 1980. There are also editorial comments on the history of the problem given there, on which our remark above is based.

meets the circumcircle at X, Y, then we have that XY is also a diameter (since it contains O). We have

$$AI \cdot IM = XI \cdot IY = (R + OI)(R - OI) = R^2 - OI^2. \qquad (5.6)$$

Let $LI \perp AB$. As $\angle LAI = \angle BAI = \angle BNM$ and $\angle ALI = \angle MBN = 90°$, we have, by the AA Similarity Rule, that $\triangle ALI \sim \triangle NBM$. So we have $AI/IL = NM/BM$, that is, $AI \cdot BM = NM \cdot IL$. Using the fact that $BM = IM$ (shown in the first paragraph above), we obtain

$$AI \cdot IM = NM \cdot IL = 2R \cdot r = 2Rr. \qquad (5.7)$$

(5.6) and (5.7) yield $OI^2 = R^2 - 2Rr$, completing the proof. $\qquad \square$

Exercise 5.19. If R, r denote the circumradius and the inradius, respectively in a triangle, then show the following:

(1) $R \geq 2r$.

(2) ($*$) $R = 2r$ if and only if the triangle is equilateral.

(3) As $(x - 1)^2 \geq 0$ for all real x, it follows by expanding the left-hand side and rearranging that for $x > 0$,

$$x + \frac{1}{x} \geq 2.$$

Taking $x := \dfrac{R}{r} > 0$, we obtain $\dfrac{R}{r} + \dfrac{r}{R} \geq 2$.

Show that in fact one has the stronger inequality $\dfrac{R}{r} + \dfrac{r}{R} \geq \dfrac{5}{2}$.

Show that the equality holds if and only if the triangle is equilateral.

Exercise 5.20. ($*$) (Circumradius in terms of the sides and the area of a triangle). Let a triangle have sides lengths a, b, c and area Δ. Then its circumradius R is given by

$$R = \frac{abc}{4\Delta}.$$

(We remark that the formula is dimensionally correct, since the left-hand side has the dimensions $[L]$ of length, while the right-hand side has the dimensions $[L]^3/[L]^2 = [L]$ as well.)

Exercise 5.21. ($*$) Let P be a point inside a triangle $\triangle ABC$ whose distances from the sides are x, y, z. Let Δ be the area, and R be the circumradius of $\triangle ABC$. Show that

$$xyz \leq \frac{2\Delta^2}{27R},$$

with equality holding if and only if P is the centroid of $\triangle ABC$.

Exercise 5.22. (∗)

(1) Prove that if three circles intersect, then their common chords are concurrent.

(2) There is a result in this context, akin to Ceva's Theorem, attributed to Hiroshi Haruki: If three circles are such that every pair intersect each other in two points, then the line segments connecting their points of intersection (with labelling as shown below) satisfy

$$\frac{BD}{DC} \cdot \frac{CE}{EA} \cdot \frac{AF}{FB} = 1.$$

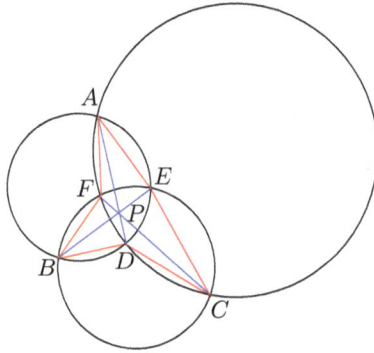

Theorem 5.16 (Butterfly Theorem). *Let M be the midpoint of the chord AB of a circle, and let CD, $C'D'$ be two other chords passing through M. Let $C'D$, CD' meet AB in P, Q, respectively. Then $PM = MQ$.*

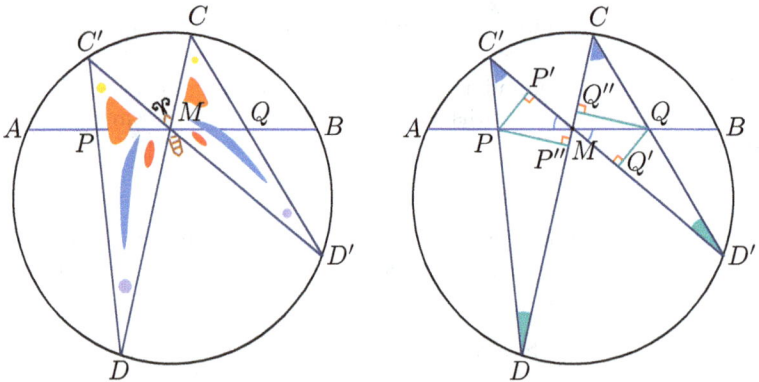

Proof. Drop perpendiculars from P to CD, $C'D'$, meeting them at P'' and P', respectively. Similarly, drop perpendiculars also from Q to CD, $C'D'$, meeting them at Q'' and Q', respectively.

$\Delta PP'M \sim \Delta QQ'M$, and so

$$\frac{PM}{QM} = \frac{PP'}{QQ'}.$$

Also, $\Delta PP''M \sim \Delta QQ''M$. So

$$\frac{PM}{QM} = \frac{PP''}{QQ''}.$$

$\Delta PP'C' \sim \Delta QQ''C$, and so

$$\frac{PP'}{QQ''} = \frac{C'P}{CQ}.$$

Finally, $\Delta PP''D \sim \Delta QQ'D'$, and

$$\frac{PP''}{QQ'} = \frac{DP}{D'Q}.$$

Consequently,

$$\begin{aligned}
\left(\frac{PM}{QM}\right)^2 &= \frac{PP'}{QQ'} \cdot \frac{PP''}{QQ''} \\
&= \frac{PP'}{QQ''} \cdot \frac{PP''}{QQ'} \\
&= \frac{C'P \cdot DP}{CQ \cdot D'Q} \\
&= \frac{AP \cdot PB}{AQ \cdot QB} \\
&= \frac{(AM - PM)(MB + PM)}{(AM + QM)(MB - QM)} \\
&= \frac{(AM - PM)(AM + PM)}{(AM + QM)(AM - QM)} \\
&= \frac{AM^2 - PM^2}{AM^2 - QM^2}.
\end{aligned}$$

With $\lambda := \dfrac{PM}{QM}$ and $\alpha := \dfrac{AM}{QM}$, the above gives

$$\lambda^2 = \left(\frac{PM}{QM}\right)^2 = \frac{AM^2 - PM^2}{AM^2 - QM^2} = \frac{\left(\dfrac{AM}{QM}\right)^2 - \left(\dfrac{PM}{QM}\right)^2}{\left(\dfrac{AM}{QM}\right)^2 - \left(\dfrac{QM}{QM}\right)^2} = \frac{\alpha^2 - \lambda^2}{\alpha^2 - 1},$$

that is, $\lambda^2 \alpha^2 - \lambda^2 = \alpha^2 - \lambda^2$. Hence $\lambda^2 = 1$, that is, $\lambda = 1$. So we have $PM = QM$. $\qquad \square$

Cyclic quadrilaterals are of course more special than mere quadrilaterals. Here is an instance of this.

Proposition 5.2. *Opposite angles in a cyclic quadrilateral are supplementary.*

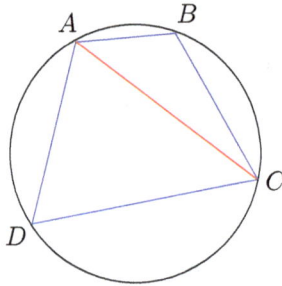

Proof. Consider a diagonal AC of the cyclic quadrilateral $ABCD$. The two angles $\angle B$ and $\angle D$ are the angles in the major and minor arc described by the chord AC, and they are supplementary by Corollary 5.5. Similarly, by considering the diagonal BD, it follows that $\angle A$, $\angle C$ are supplementary too. (Or by using the supplementarity of $\angle B$ and $\angle D$ just shown, and the fact that all four angles add up to $360°$.) □

Exercise 5.23. Show that if a pair of opposite angles in a quadrilateral are supplementary, then the quadrilateral is cyclic.

Exercise 5.24. What can you say about a cyclic parallelogram?

Exercise 5.25. Show that any cyclic trapezium must have equal nonparallel sides.

Exercise 5.26. Prove that the intersection points of the adjacent angle bisectors in a quadrilateral $ABCD$ form a cyclic quadrilateral $A'B'C'D'$.

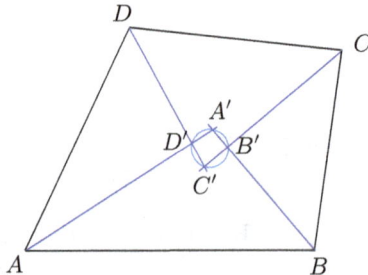

Exercise 5.27. (Symmedians). A *symmedian* of a triangle is the reflection of the median through the corresponding angle bisector. See the picture below.

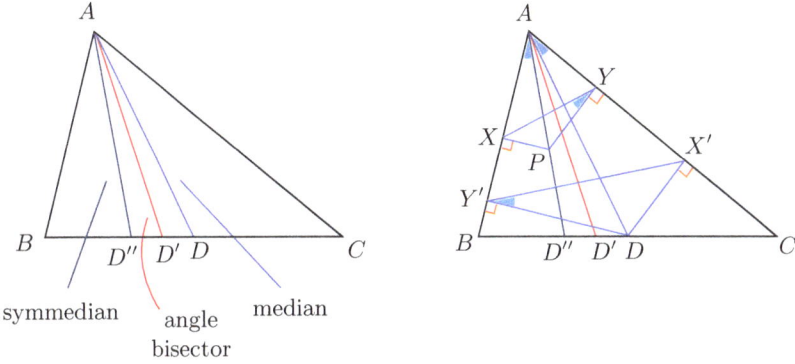

symmedian   angle   median
bisector

Let a, b, c be the lengths of the sides BC, CA, AB, respectively.

(1) Let P be any point on AD'', and drop perpendiculars from P to the sides AB, AC, meeting them at X, Y, respectively. Drop perpendiculars to AB, AC from D, meeting them in Y', X', respectively. By considering the cyclic quadrilaterals $AXPY$ and $AX'DY'$ shown in the picture above, prove that $\triangle XPY \sim \triangle X'DY'$.

Furthermore, by considering the equality of the areas of $\triangle ABD$ and $\triangle ADC$, show that the distances x, y of P to the sides AB and AC are in proportion to the side lengths, that is, $x : y = c : b$.

(2) Show that the symmedian AD'' divides the side BC in the ratio $c^2 : b^2$.

(3) Prove that the symmedians are concurrent. Their point of concurrency is called the *Lemoine Point*.

(4) Let P be any point inside the triangle, and consider the problem of minimizing the sum of the squares of the distances of P to the three sides. Show that the Lemoine point does the job.

Theorem 5.17 (Bramhagupta's Formula). *If a cyclic quadrilateral has sides of lengths a, b, c, d, then its area is given by*

$$\sqrt{(s - a)(s - b)(s - c)(s - d)}$$

where $s := \dfrac{a + b + c + d}{2}$ *is the semiperimeter.*

Proof. If both pairs of opposite sides are parallel, then the cyclic quadrilateral is a rectangle, and so $a = c$ and $b = d$, giving $s = a + b$ and $\sqrt{(s - a)(s - b)(s - c)(s - d)} = ab$, which is indeed the area of the rectangle. So we may assume in the sequel that at least one pair of opposite sides is nonparallel. Hence without loss of generality, we may consider the following picture, where the produced sides AB and CD meet at P, and $BP = x$, while $DP = y$.

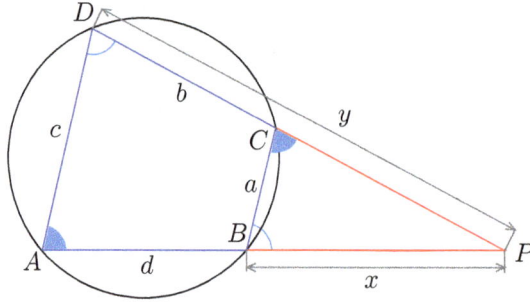

As $\angle BCP + \angle BCD = 180° = \angle BCD + \angle BAD$, we have $\angle BCP = \angle PAD$. Similarly, $\angle ADP = \angle CBP$. By the AA Similarity Rule, $\Delta BCP \sim \Delta DAP$. Let the ratio of the sides of the smaller triangle to the bigger triangle be $\lambda \ (< 1)$. Let F_1, F_2, F_3, F_4 be the four factors in Heron's Formula for the area of ΔAPD. Then the area of the quadrilateral $ABCD$ is given by

$$
\begin{aligned}
\text{Area}(ABCD) &= \Delta APD - \Delta BCP \\
&= \sqrt{F_1 F_2 F_3 F_4} - \sqrt{(\lambda F_1)(\lambda F_2)(\lambda F_3)(\lambda F_4)} \\
&= (1 - \lambda^2)\sqrt{F_1 F_2 F_3 F_4} = (1 - \lambda)(1 + \lambda)\sqrt{F_1 F_2 F_3 F_4} \\
&= \sqrt{(1 - \lambda)F_1(1 - \lambda)F_2(1 + \lambda)F_3(1 + \lambda)F_4} \\
&= \sqrt{(F_1 - \lambda F_1)(F_2 - \lambda F_2)(F_3 + \lambda F_3)(F_4 + \lambda F_4)}.
\end{aligned}
$$

But

$$
F_1 - \lambda F_1 = \frac{c + y + (d + x)}{2} - \frac{(y - b) + a + x}{2} = \frac{b + c + d - a}{2} = s - a,
$$

$$
F_2 - \lambda F_2 = \frac{y + (d + x) - c}{2} - \frac{x + (y - b) - a}{2} = \frac{a + b + d - c}{2} = s - c,
$$

$$
F_3 + \lambda F_3 = \frac{(d + x) + c - y}{2} + \frac{(y - b) + a - x}{2} = \frac{a + c + d - b}{2} = s - b,
$$

$$
F_4 + \lambda F_4 = \frac{c + y - (d + x)}{2} + \frac{a + x - (y - b)}{2} = \frac{a + b + c - d}{2} = s - d.
$$

Consequently, the area of the cyclic quadrilateral $ABCD$ is

$$
\begin{aligned}
&\sqrt{(F_1 - \lambda F_1)(F_2 - \lambda F_2)(F_3 + \lambda F_3)(F_4 + \lambda F_4)} \\
&= \sqrt{(s - a)(s - b)(s - c)(s - d)}.
\end{aligned}
$$

$\square$

Exercise 5.28. What happens when d diminishes to 0?

Exercise 5.29. $(*)^{15}$ A quadrilateral inscribes a circle and also circumscribes one. If the sides of the quadrilateral have lengths a, b, c, d, then show that its area is $\sqrt{abcd}$.

Exercise 5.30. $(*)$ In a triangle $\triangle ABC$, the line segment joining the orthocenter O and the midpoint D of the side BC meets the circumcircle of $\triangle ABC$ at the point O'. Show that D is the midpoint of OO'.

Given a circle, any line which intersects the circle at two distinct points is called a *secant* to the circle. The following result complements the result from Theorem 5.14.

Theorem 5.18. *Let P be any point outside a given circle $C(O, r)$, and let ℓ be any secant to the circle passing through P, and intersecting the circle at the points A_ℓ, B_ℓ. Then the product $PA_\ell \cdot PB_\ell$ is a constant (depending only on $C(O, r)$ and P).*

Proof. Look at the picture below, where ℓ, ℓ' are any two secants passing through the point P. We want to show that $PA_\ell \cdot PB_\ell = PA_{\ell'} \cdot PB_{\ell'}$.

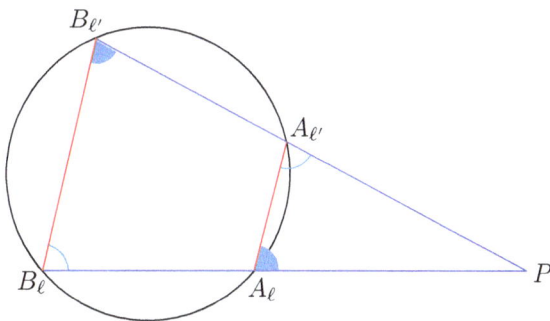

Since $\angle A_{\ell'} A_\ell P + \angle A_{\ell'} A_\ell B_\ell = 180° = \angle B_\ell B_{\ell'} P + \angle A_{\ell'} A_\ell B_\ell$, it follows that

$$\angle A_{\ell'} A_\ell P = \angle B_\ell B_{\ell'} P.$$

Similarly, $\angle A_\ell A_{\ell'} P = \angle P B_\ell B_{\ell'}$. Thus by the AA Similarity Rule, we conclude that $\triangle P A_\ell A_{\ell'} \sim \triangle P B_{\ell'} B_\ell$. So their sides are proportional, that is,

$$\frac{PA_\ell}{PB_{\ell'}} = \frac{PA_{\ell'}}{PB_\ell}.$$

Hence $PA_\ell \cdot PB_\ell = PA_{\ell'} \cdot PB_{\ell'}$, as required. $\qquad \square$

---

[15]This exercise can wait until Theorem 5.23 has been covered.

Exercise 5.31. A circle intersects the sides AB, BC, CA at the points D, D', E, E', F, F', as shown in the picture. If AD, BE, CF are concurrent, then show that AD', BE', CF' are concurrent as well.

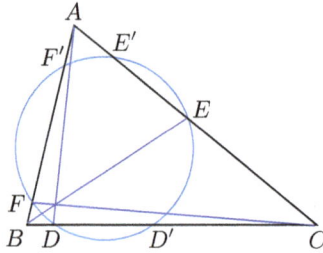

Arc length of a circular arc and the area of a sector

Consider a circle of radius r. Clearly a semicircular arc of this circle has arc length equal to half the circumference, $(2\pi r)/2 = \pi r$, and if we divide the subtended angle of measure $180°$ at the center into 180 equal parts, then the circular arc is divided into 180 circular arcs, all of which are congruent to each other. Hence the arc length of a circular arc which subtends an angle of $1°$ at the center is $(\pi r)/180$. Consequently the arc length of a circular arc which subtends an angle of $n°$ at the center, where n is a natural number such that $0 \le n \le 180$ is $n \cdot \pi r/180 = n\pi r/180$.

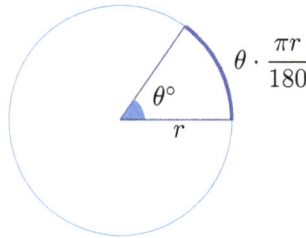

If we divide the angle $n°$ this circular arc subtends at the center into d equal parts, where d is any natural number, then we obtain d circular arcs, each of which has arc length equal to

$$\frac{(n\pi r)/180}{d} = \frac{n}{d} \cdot \frac{\pi r}{180}.$$

Thus the length of a circular arc which subtends a rational angle $\theta°$ at the center is $\theta \cdot \pi r/180$. As all real numbers can be approximated by rationals to an arbitrary degree of accuracy, we conclude that if a circular arc subtends an angle of $\theta°$ (where θ is any real number, rational or irrational), at the center of the circle of radius r, then its arc length is given by $\theta \cdot \pi r/180$.

Similar considerations show that the area of a circular arc which subtends an angle of $\theta°$ at the center is $(\theta/360) \cdot \pi r^2$.

Exercise 5.32. (∗) (Copernicus's Theorem) Consider a circle of a radius $2r$, and suppose that there is a smaller circle of radius r inside the bigger circle which rolls within it without slipping. What is the trajectory of a point P on the smaller circle? (That is, imagining a tiny ribbon tied to the rim of the smaller circle, what is the path traced out by this ribbon?)

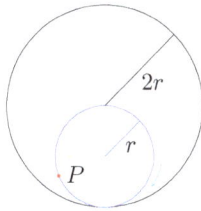

Newton's inverse square law deduction from Kepler's laws

We will derive Newton's inverse square law for gravitational force from Kepler's laws of planetary motion. Kepler's laws were listed on page 66. Here we will consider the case when a planet is moving in a circular orbit with the Sun at the center of the orbit. (A circle is of course a special case of an ellipse, where the two foci are coincident, and so a circular orbit is a special case of the elliptical orbits dictated by Kepler's First Law.)

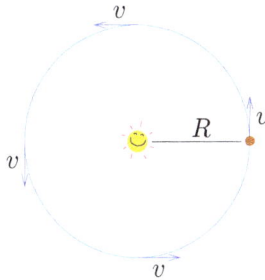

Kepler's Second Law, and the formula for the circular sector area in terms of the central angle show that the angular speed must be a constant. The formula for the circular arc length now allows us to conclude that the speed $v(t)$ at time t must not change with t, and is a constant, say v. The time taken to finish one revolution around the Sun is thus $T = 2\pi R/v$.

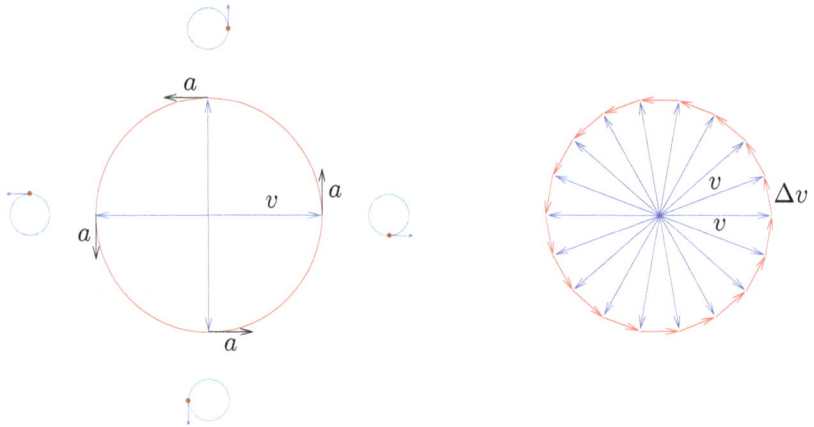

What is the circular acceleration? Note first that the velocity vector also rotates around the circle (and the time it takes to do so is the same as T). We note that as the situation is rotationally symmetric, the circular acceleration must have the same magnitude, say a, everywhere. In the picture on the right above, by looking at a tiny isosceles triangle formed by two velocities at time instances ΔT apart, we also note that the angle made by the side Δv with any of the two long sides, which include a tiny angle, is close to $90°$, from which we see that the acceleration vector is perpendicular to the velocity vector at any time instant. This means that the acceleration at any point of the circular orbit is always directed along the radius towards the center of the circle. In order to find the magnitude of the acceleration, we will make use of the fact that the velocity vector takes time T in order to rotate once around the circle. Now the acceleration magnitude is the limiting value of the ratio

$$\frac{\Delta v}{\Delta T}$$

as ΔT diminishes to 0. But thanks to the fact that the acceleration magnitude is the same everywhere, calculating the limiting value of this ratio is the same as taking the sum of the lengths of the sides of the polygon in the right-hand side picture shown above, and dividing it by the total time T the velocity vector needs to go around once the vertices of the polygon, as the number of the sides of the polygon goes to infinity. But this is just the ratio of the circumference of the circle of radius v to the time T. Hence the acceleration magnitude must be

$$a = \frac{2\pi v}{T} = \frac{4\pi^2 R}{T^2}.$$

By Newton's Second Law of Motion, we must have that the magnitude F of the force, being proportional to the magnitude of the acceleration, is

$$F \propto \frac{R}{T^2}.$$

Finally, from Kepler's Third Law of planetary motion, we know that the square of the orbital period T of a planet is proportional to the cube of the semi-major axis of its (elliptical) orbit, and in the case of a circular orbit, this is just the radius R of the circle. Hence

$$F \propto \frac{R}{T^2} \propto \frac{R}{R^3} = \frac{1}{R^2}.$$

Moreover, since the force is a scalar multiple of the acceleration vector ($\overrightarrow{F} = m\overrightarrow{a}$), it has the same direction as the acceleration (which we had seen is directed radially inward). Thus we have obtained Newton's inverse square law for gravitation!

5.3 Tangent line to a circle

Given a circle and a line in the plane, there are only three possibilities:

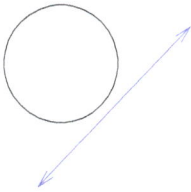

1° The line does not intersect the circle.

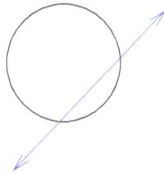

2° The line intersects the circle in exactly two points.

3° The line touches the circle at exactly one point.

Indeed, if a circle intersects a line in A and B, then the center of the circle lies on the perpendicular bisector of the chord AB. If there is a *third* intersection point C, then the center of the circle must also lie on the perpendicular bisector of BC. But these two perpendicular bisectors are distinct parallel lines, and cannot have a point in common.

In the last case when the line touches the circle at exactly one point, we say that the line is *tangent* to the circle, and call it a *tangent line*, while the common point between the tangent line and the circle is referred to as the *point of tangency*.

Theorem 5.19. *If a line is tangent to a circle, then the line joining the center of the circle to the point of tangency is perpendicular to the tangent line.*

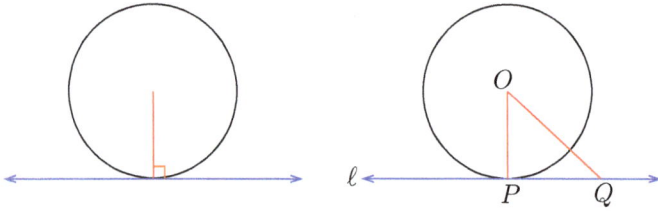

Proof. Let ℓ be the tangent line. Let O be the center of the circle, and P be the point of tangency. If $Q \neq P$ is any point on ℓ, then Q must lie outside the circle. (Q can't be on the circle, since ℓ, being a tangent, ought to meet the circle at *one* point alone. Also Q can't be inside the circle either, because then PQ extended would meet the circle at another point, and so again this contradicts the fact that ℓ is a tangent line!) But then $OQ >$ radius of the circle $= OP$. As this holds for all $Q \neq P$ on ℓ, it follows that P is the point on ℓ which is closest to O. But by Theorem 3.10, we then conclude that $OP \perp \ell$.   □

Theorem 5.20. *If T is a point on the circle $C(O,r)$, P is a point outside the circle $C(O,r)$, and $\angle OTP = 90°$, then PT is a tangent to $C(O,r)$ with point of tangency T.*

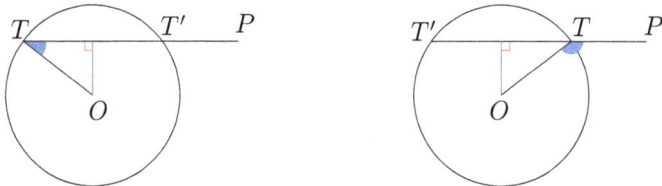

Proof. As T is a point on the circle, the line ℓ through P, T does meet the circle. Suppose that ℓ is not tangential to $C(O,r)$. Then ℓ meets $C(O,r)$ at another point $T' \neq T$. The perpendicular bisector of TT' contains the center O of the circle. From the two pictures above, we see that $\angle OTP$ is either $< 90°$ (left) or $> 90°$ (right), a contradiction.   □

Construction 5.3 (Tangent to a circle from a point). *Given a circle $C(O,r)$, its center O, and a point P outside the circle, construct a tangent line to $C(O,r)$ which passes through P.*

Join P to O, and construct the midpoint M of the line segment OP. With M as center, and radius OM, draw a circle $C(M, OM)$. The circle $C(M, OM)$ will intersect $C(O, r)$ in two points. Join P to any one of these two points, say T. Then PT is the required tangent line.

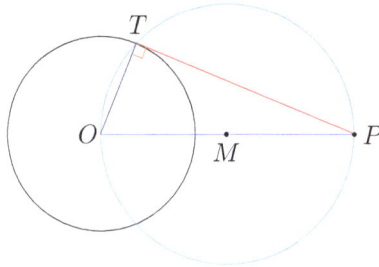

Rationale: $\angle OTP = 90°$, since it is the angle subtended by the diameter OP at the point T on the circle $C(M, OM)$. The radius OT is perpendicular to the line PT, and hence PT must be the tangent line. $\diamond$

Theorem 5.21. *If ℓ is a tangent line to a circle with point of tangency T, and $S \neq T$ is any point on the circle, then the angles subtended by the chord ST at any point on the major/minor arc equal the respective angles that ST makes with the line ℓ.*

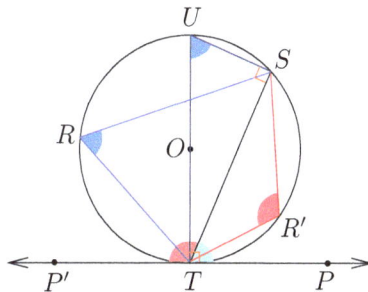

Proof. By referring to the picture above, let TU be the diameter through T, and R, R' be any two points on the major/minor arcs described by the chord ST. Then we have $\angle TSU = 90°$ since it is the angle subtended by a diameter, and $OT \perp TP$, since the radius is perpendicular to the tangent line at the point of tangency. So we have

$$\angle STP = 90° - \angle UTS = 90° - (90° - \angle TUS) = \angle TUS = \angle TRS.$$

Moreover, since the angles $\angle TRS$ and $\angle TR'S$ are supplementary, it follows that $\angle P'TS = 180° - \angle STP = 180° - \angle TRS = \angle TR'S.$ $\square$

Theorem 5.22. *If a ray $\overrightarrow{TP}$ is drawn through the end point T of a chord ST of a circle so that the angle $\angle STP$ formed with the chord equals the angle subtended by the chord in the arc described by ST which lies on the side of ST not containing the point P, then TP is tangential to the circle with point of tangency T.*

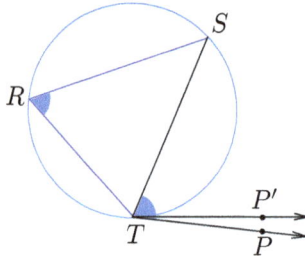

Proof. Suppose that $\overrightarrow{TP'}$ is tangential to the given circle with point of tangency T, and that $\overrightarrow{TP'}$ is on the same side of ST as the point P. Let R be any point on the circle which is on the side of ST which does not contain P. Then it is given that $\angle TRS = \angle STP$. But from the previous theorem, $\angle TRS = \angle STP'$. Hence $\angle STP = \angle STP'$, and so the ray $\overrightarrow{TP'}$ coincides with the ray $\overrightarrow{TP}$.   □

Exercise 5.33. Given a circle with an unknown center, and a point T on the circle, draw a tangent line to the circle with point of tangency T.

Exercise 5.34. (∗) Construct $\triangle ABC$ given a line segment with length BC, an angle with measure $\angle A$, and a line segment of length equal to the median AD.

Exercise 5.35. (∗) Construct $\triangle ABC$ given a line segment with length BC, an angle with measure $\angle A$, and a line segment of length equal to the altitude AD.

Exercise 5.36. (Estimation[16] of sizes of the Sun, the Moon, and their distances to the Earth). Let

$$R_{\mathbb{C}} := \text{radius of the Moon,}$$
$$R_{\odot} := \text{radius of the Sun,}$$
$$d_{\mathbb{C}} := \text{distance between the Earth and the Moon,}$$
$$d_{\odot} := \text{distance between the Earth and the Sun.}$$

Step 1: One can estimate the angle between a "half moon" (or first quarter moon) and the new moon by measurement, and this turns out to be about $89.85°$. We expect $d_{\odot}$ to be much larger than $d_{\mathbb{C}}$. So $d_{\mathbb{C}}$ is approximately equal to the circular

---

[16]Following the Greek astronomer Aristarchus (280-240 BC).

arc length of a circular arc of radius $d_\odot$ subtending an angle of $90° - 89.85° = 0.15°$ at the center. Thus $d_{\mathbb{C}} / d_\odot \approx 0.15 \cdot (\pi/180) \approx 1/381$. See the following picture.

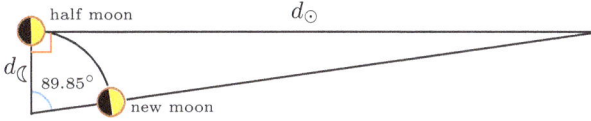

Step 2: During a solar eclipse, one notices that the Moon almost covers the Sun. So based on the picture below, $d_{\mathbb{C}} / d_\odot = R_{\mathbb{C}} / R_\odot$.

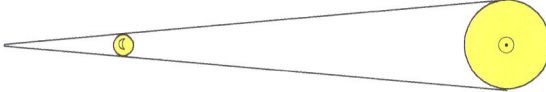

Step 3: We can estimate the relative size of the Moon with respect to the Earth as follows. First of all, we can estimate the time $T_{\mathbb{C}}$ that the Moon takes to sweep its own diameter in the sky. This can be done by looking at the relative position of the Moon relative to some distant stars.

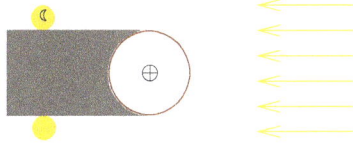

During a total lunar eclipse, we can also estimate the time $T_\oplus$ it takes for the Moon to pass through the shadow of the Earth. Assume that the shadow is cylindrical, of the same radius as that of the Earth. Then $R_{\mathbb{C}} / R_\oplus \approx T_{\mathbb{C}} / T_\oplus$.

Step 4: Knowing that it takes 28 days for the Moon to go around the Earth, we can estimate the circumference of Moon's trajectory around the Earth, and so $2\pi d_{\mathbb{C}} = (28 \times 24 \text{ hours}) \cdot (2 \times R_{\mathbb{C}} / T_{\mathbb{C}} \text{ km/hours})$. Suppose that the time the Moon takes to sweep its own diameter is 0.93 hours, and during a lunar eclipse, the time the Moon takes to pass Earth's shadow is about 3.39 hours. Take $R_\oplus = 6400$ km. Estimate $R_{\mathbb{C}}, d_{\mathbb{C}}, R_\odot$ and $d_\odot$.

Exercise 5.37. (Honey, where should we sit?) On Valentine's Day, a couple decide to go to the cinema. They decide to sit in the middle of a row, and want to choose the row along the sloping floor that maximizes the viewing angle θ of the screen AB, as shown below. Which point "row" P achieves this?

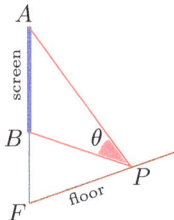

Theorem 5.23. *If P is a point outside a circle with center O and T_1, T_2 are the points of tangency of tangents to the circle drawn from P, then:*

(1) *$PT_1 = PT_2$ and*
(2) *the line OP bisects $\angle T_1 P T_2$.*

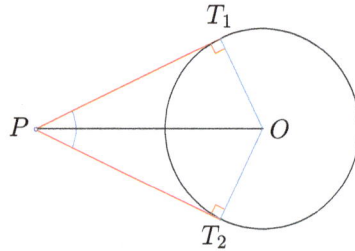

The first part of the result is clear from Theorem 5.18: indeed, if we imagine two secants ℓ_1, ℓ_2 moving in opposite directions, approaching the tangent lines PT_1, PT_2, then we expect that $PA_{\ell_1} \cdot PB_{\ell_1}$ approaches PT_1^2, while $PA_{\ell_2} \cdot PB_{\ell_2}$ approaches PT_2^2, and as $PA_{\ell_1} \cdot PB_{\ell_1} = PA_{\ell_2} \cdot PB_{\ell_2}$ for all positions of ℓ_1, ℓ_2, it follows that $PT_1^2 = PT_2^2$, and so $PT_1 = PT_2$.

Proof. $OT_1 \perp PT_1$ and $OT_2 \perp PT_2$. Join PO. In the right angled triangles $\triangle PT_1O$ and $\triangle PT_2O$, PO is a common side, $OT_1 = OT_2$ (both equal to the radius of the circle) and $\angle PT_1O = \angle PT_2O = 90°$. By the RHS Congruency Rule, $\triangle PT_1O \simeq \triangle PT_2O$. Hence $PT_1 = PT_2$ (CPCT) and $\angle T_1PO = \angle T_2PO$ (CPCT). $\qquad\square$

Exercise 5.38. A right triangle has an inscribed circle of radius r and a circumscribed circle of radius R. Show that the average of the two nonhypotenuse sides equals $R + r$.

Exercise 5.39. From a point P outside a circle, tangents PT_1, PT_2 are drawn with points of tangency T_1 and T_2. Show that $\angle T_1PT_2 = 2\angle OT_1T_2$.

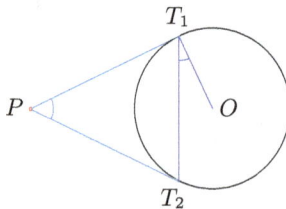

Corollary 5.7. *If the incircle of a triangle touches the sides AB, BC, CA at X, Y, Z, then AX, BY, CZ are concurrent.*

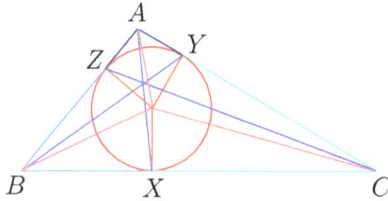

Proof. As $BX = BZ$, $CX = CY$, $AY = AZ$, it follows that

$$\frac{BX}{XC} \cdot \frac{CY}{YA} \cdot \frac{AZ}{ZB} = 1.$$

By the converse of Ceva's Theorem, AX, BY, CZ are concurrent. □

Theorem 5.24. *Let P be any point outside a given circle $C(O, r)$. Suppose that ℓ is any secant to the circle passing through P intersecting the circle at the points A, B. Let PT be a tangent to the circle with point of tangency T. Then $PT^2 = PA \cdot PB$.*

Again this is evident from Theorem 5.18: indeed, if we imagine two secants ℓ, ℓ', passing through P, where ℓ is fixed, while ℓ' is being rotated about the point P so that it is approaching the tangent line PT, then for all positions of ℓ', we have that $PA \cdot PB = PA_\ell \cdot PB_\ell = PA_{\ell'} \cdot PB_{\ell'}$, and we expect the last expression approaches PT^2 as ℓ' approaches the tangent line PT.

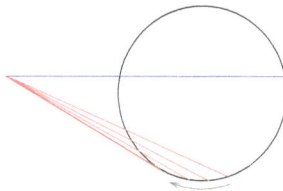

Proof. Join OP, and let it meet the circle at A', and when extended, at B'. See the following picture.

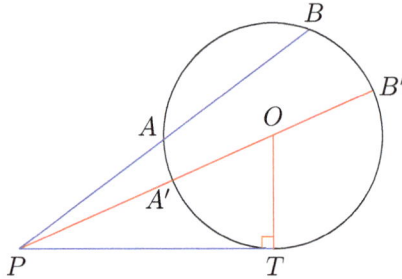

OA', OT, OB' are all equal to the radius r. By Theorem 5.18, we have that $PA' \cdot PB' = PA \cdot PB$. Using Pythagoras's Theorem in the right angled triangle $\triangle PTO$, we obtain

$$PT^2 = PO^2 - OT^2 = (PO - OT)(PO + OT)$$
$$= (PO - OA')(PO + OB') = PA' \cdot PB'$$
$$= PA \cdot PB.$$

$\square$

Construction 5.4 (Tangent to a circle from a point). *Given a circle C, but with no knowledge of its center, and a point P outside it, construct a tangent line to the circle C which passes through P.*

We could first find the center using Corollary 5.1 as follows: take any three points P, Q, R on the circle, draw the perpendicular bisectors of the chords PQ, QR; then the point of intersection of these perpendicular bisectors is the center of the circle C. Then we use Construction 5.3 to draw the required tangent line to C through P.

Alternatively, we can also perform the required construction (without determining the center) in light of Theorem 5.24 (and by noting that in Exercise 5.7 we learnt a method for constructing the geometric mean of two given numbers). So we could do the following. Through P draw any line which cuts C at two points, A, B. Then we know PA, PB, and hence we can construct a line segment with length $d := \sqrt{PA \cdot PB}$. With P as center and radius d, draw a circle $C(P, d)$. Then $C(P, d)$ will meet the given circle C at two points. Take any one of these two points, and call it T. Join P to T. Then PT is the required tangent line. $\diamond$

Exercise 5.40. Two points A, B are given on the same side of a line ℓ. Construct a circle which passes through A, B such that ℓ is tangential to it.

Exercise 5.41. (An entry from Ramanujan's[17] Notebook). Entry 1 of the unorganized material of Ramanujan's second and third notebooks contains the following picture, where $\triangle ABC$ is a right angled triangle, and the two circular arcs shown have centers B and C. See the picture on the left below. Among the three formulas that Ramanujan noted, the first one is $PQ^2 = 2BP \cdot QC$. Show this.

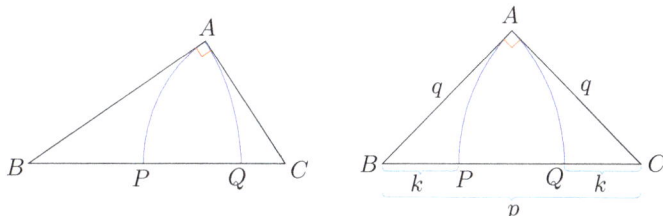

In fact, the above formula delivers yet another geometric proof that $\sqrt{2}$ is irrational. Indeed, suppose that $\sqrt{2} = p/q$ for some positive integers p, q, and suppose that q is the smallest such positive integer. As $1 < \sqrt{2} < 2$, we obtain that $q < p < 2q$. Write $p = q + k$, with the integer k satisfying $1 \le k < q$. As $p^2 = 2q^2$, it follows, from the geometric observation above (see the picture on the right above), that $(q - k)^2 = (p - k - k)^2 = PQ^2 = 2BP \cdot QC = 2 \cdot k \cdot k = 2k^2$, and so $\sqrt{2} = (q - k)/k$, contradicting our choice of q.

5.4   An excursion in inversion

Let $C(O, r)$ be a circle in a plane $\mathbb{P}$. Given a point $P \in \mathbb{P} \setminus O$, we can look at the ray $\overrightarrow{OP}$, and find the unique point P' on it which satisfies $OP \cdot OP' = r^2$. In other words, P' is the point whose distance to O is $OP' = r^2/OP$. Note that $OP' > 0$ so that $P' \ne O$, and so we get a map $P \mapsto P' : \mathbb{P} \setminus \{O\} \to \mathbb{P} \setminus \{O\}$, which we call *inversion* (with respect to the circle $C(O, r)$).

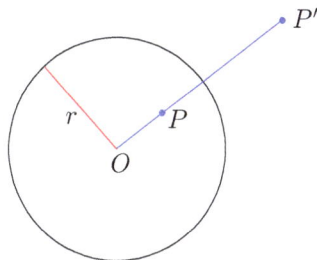

[17]Srinivas Ramanujan (1887-1920) was an Indian mathematician who made extraordinary contributions to analytic number theory. Ramanujan recorded many of his results in four notebooks, without any derivations.

It is clear that if P' is the inversion of P, then the inversion of P' is P. It will be convenient to denote the inversion map by $I : \mathbb{P} \setminus \{O\} \to \mathbb{P} \setminus \{O\}$, and the image of a point $P \in \mathbb{P} \setminus \{O\}$ will be denoted by $I(P)$.

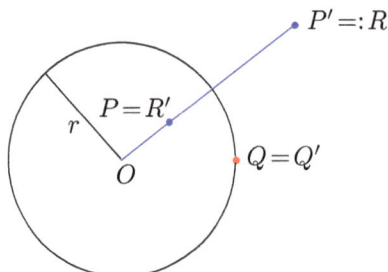

Why the name "inversion"? Points *inside* the circle $C(O, r)$ are mapped *outside* it, and vice versa, while points on the circle stay put.

Also, circle inversion is a generalization of the usual reflection in a straight line. Indeed, if we think of a point P inside $C(O, r)$ at a distance d from $C(O, r)$, and whose image P' is at a distance d' from $C(O, r)$, then $OP \cdot OP' = r^2$ becomes $(r - d)(r + d') = r^2$, that is,

$$\frac{1}{d} - \frac{1}{d'} = \frac{1}{r}.$$

Now if we imagine r becoming larger and larger, that is, imagine O moving towards the left in the picture on the left below, while keeping the point Q on the circle, as well as the point P fixed, then we see that the circular arc looks more and more like a vertical line ℓ, and the equation above gives $d = d'$, reducing the map $P \mapsto P'$ to reflection in ℓ.

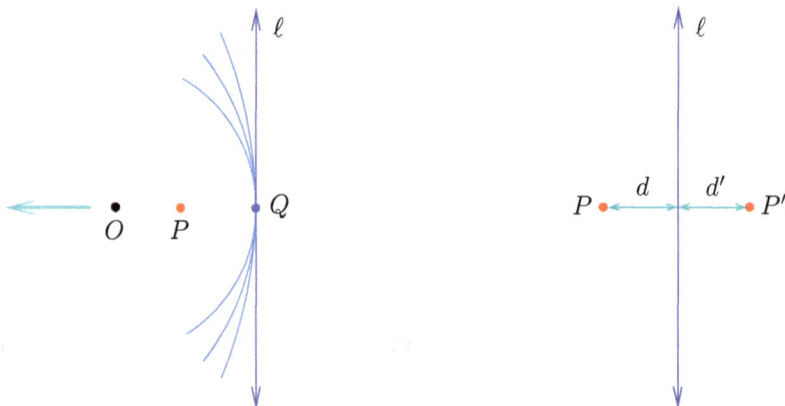

How do we construct the image of a given point under the inversion map?

Construction 5.5 (Inversion of a point in a circle). *Given* $C(O,r)$ *and* $P \neq O$*, construct* $I(P)$.

We consider the three possible cases that P lies on/inside/outside $C(O,r)$:

1° P lies on $C(O,r)$. Then $I(P) = P$.

2° P lies inside $C(O,r)$. Join O to P, and erect a perpendicular at P to OP, meeting the circle at Q, say. Join O to Q, and erect a perpendicular at Q to OQ, meeting OP extended at P'. Then P' is the required point.

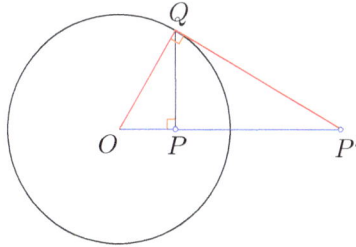

Indeed, $\triangle OPQ \sim \triangle OQP'$, and so $OP/OQ = OQ/OP'$, that is, we have $OP \cdot OP' = OQ^2 = r^2$. Thus $P' = I(P)$.

3° P lies outside $C(O,r)$. Join OP, and draw a circle with OP as diameter, intersecting $C(O,r)$ in, say, Q. Drop a perpendicular from Q onto OP, meeting OP at P'. Then P' is the required point.

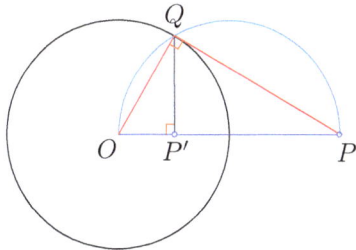

Indeed, $\triangle OP'Q \sim \triangle OQP$, and so $OP'/OQ = OQ/OP$, that is, we have $OP' \cdot OP = OQ^2 = r^2$. Thus $P' = I(P)$. ◇

The images of pairs of points form similar triangles with the center.

Proposition 5.3. *Let* $C(O,r)$ *be a circle. Suppose that* A, B *are points inside* $C(O,r)$ *different from* O*, and their images under inversion in* $C(O,r)$

are A', B', respectively. Then $\triangle AOB \sim \triangle B'OA'$.

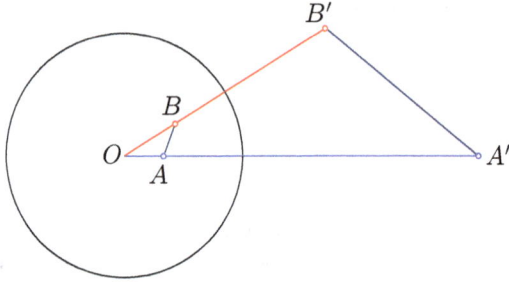

Proof. In the triangles $\triangle AOB \sim \triangle B'OA'$, $\angle AOB = \angle B'OA'$ (common), and $OA \cdot OA' = r^2 = OB \cdot OB'$, that is,

$$\frac{OA}{OB'} = \frac{OB}{OA'}.$$

Hence by the SAS Similarity Rule, $\triangle AOB \sim \triangle B'OA'$.   □

Conformality

In ordinary reflection in a line, the distances and angles are preserved, and the image is congruent to the original figure. However, now in circle inversion, the figures are distorted, but angles are preserved, and we will show this in the next paragraph.

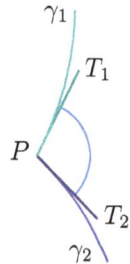

Imagine two intersecting "smooth" curves γ_1 and γ_2 at a point P. What is the angle between them? Well, by "smooth" curves, we mean that the curves can be approximated by straight lines in the vicinity of P, that is, the curves have "tangents" PT_1 and PT_2 near P. So near P, there is very little difference between γ_1 and its tangential approximation PT_1, and similarly for γ_2 and PT_2. Thus we can define the angle between γ_1 and γ_2 to be the angle between PT_1 and PT_2.

If T_1', T_2', P' are the images under inversion in the circle $C(O, r)$ of the three points T_1, T_2, P, respectively, then using Proposition 5.3, it follows that $\Delta OP'T_1' \sim \Delta OT_1P$ and also that $\Delta OP'T_2' \sim \Delta OT_2P$. These pairs of similar triangles allow us to conclude that $\angle OP'T_1' = \angle OT_1P$ and also that $\angle OP'T_2' = \angle OT_2P$.

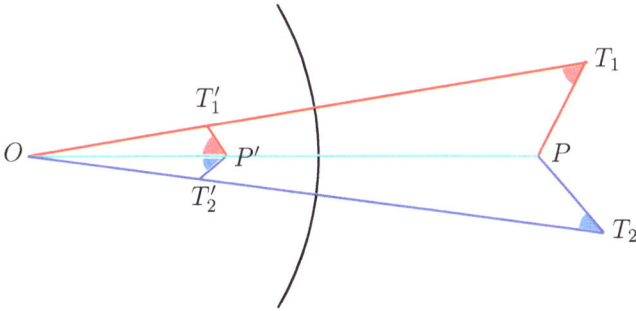

Now the images of the two curves γ_1 and γ_2 under the inversion map will be two other intersecting curves with point of intersection P', and near P', we can replace $I(\gamma_1)$ and $I(\gamma_2)$ by the images $P'T_1'$ and $P'T_2'$. Thus the angle between the images of the curves is the angle between $T_1'P'$ and $T_2'P'$, namely

$$\angle T_1'P'T_2' = \angle OP'T_1' + \angle OP'T_2'$$
$$= \angle OT_1P + \angle OT_2P$$
$$= \angle T_1PT_2 - (\angle T_1OT_2).$$

But as T_1, T_2 are both near P, it follows that $\angle T_1OT_2 \approx 0$. Thus eventually we have $\angle T_1'P'T_2' = \angle T_1PT_2$, and so the angle between curves is preserved. This angle preservation property between curves is called "conformality".

Using Proposition 5.3, we can show that images of circles inside $C(O, r)$ which don't pass through O are circles! This is analogous to what happens in the usual reflection in a line.

Proposition 5.4. *Let $C(O, r)$ be a circle, and $\mathcal{C}$ be a circle inside $C(O, r)$, not passing through O, with the points A, B as the diameter. Let A', B' be the images under inversion in $C(O, r)$ of the points A, B, respectively. Then the image of $\mathcal{C}$ is the circle $\mathcal{C}'$ having the diameter $A'B'$.*

Proof. This is essentially a "proof without words" by meditating on the following two pictures.

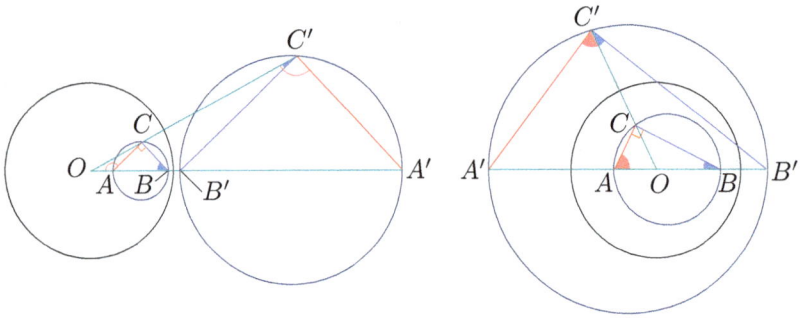

Let C be any point on the circle $\mathcal{C}$, and let C' be the image of C under inversion in $C(O,r)$. Thanks to the facts that $\triangle AOC \sim \triangle C'OA'$ and $\triangle BOC \sim \triangle C'OB'$, we have $\angle A'C'O = \angle CAO$ and $\angle B'C'O = \angle CBO$. But as $\angle ACB = 90°$, it now follows (by subtraction of the previous two equations for the left picture, and by addition for the right picture) that $\angle B'C'A' = 90°$ too. Hence C' lies on the circle $\mathcal{C}'$ with diamater $A'B'$. □

It is natural to ask what happens if the given circle $\mathcal{C}$ actually passes through O, that is, what if A coincides with O? In that case, of course I won't map the point A to anything in the plane $\mathbb{P}$, but it turns out that the image of $\mathcal{C} \setminus \{O\}$ is a line passing through B', perpendicular to OB'. This is intuitively clear from the left picture above: Let us see what happens to $\mathcal{C}'$ as the point A moves towards O, while the point B is kept fixed. Then we see that for the circle $\mathcal{C}'$, the point B' stays fixed, while the point A' goes farther and farther away from B', and the portion of the circle $\mathcal{C}'$ near the point B' "straightens out"! Thus we expect the following:

Proposition 5.5. *Let $C(O,r)$ be a circle, and $\mathcal{C}$ be a circle inside $C(O,r)$ passing through O, with OB as the diameter. Let B' be the image under inversion in $C(O,r)$ of the point B. Then the image of $\mathcal{C} \setminus \{O\}$ is the straight line perpendicular to OB', passing through B'.*

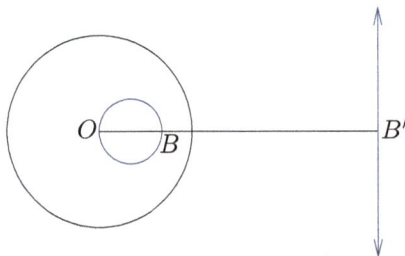

Proof. Let C be any point on the circle $\mathcal{C}$ except the point O, and extend OC to meet the line through B' perpendicular to OB' at the point C'.

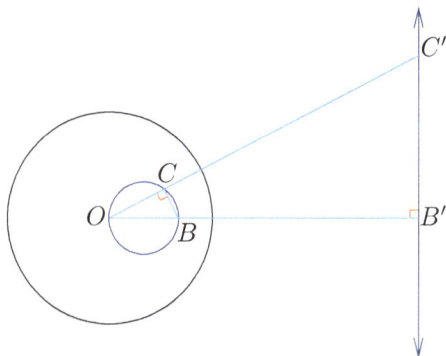

As we know that $\angle OCB = 90° = \angle OB'C'$, and since $\angle BOC = \angle C'OB'$ (common), it follows that $\triangle BOC \sim \triangle C'OB'$. Thus

$$\frac{OB}{OC'} = \frac{OC}{OB'},$$

that is, $OC \cdot OC' = OB \cdot OB' = r^2$. Consequently C' is the image of C under inversion in the circle $C(O, r)$.

Vice versa, it is also clear, using the property that $I(I(P)) = P$ for every point $P \in \mathbb{P} \setminus \{O\}$, that if we start with any line ℓ outside C, and drop a perpendicular to ℓ from the center O, meeting ℓ in B', then the image of ℓ under the map I is a circle passing through O and through $I(B') =: B$. $\square$

Inversion in the circle can be used to one's advantage, as demonstrated in the instances below.

Ptolemy's Theorem redux

Recall that Ptolemy's Theorem states that in any cyclic quadrilateral $ABCD$, $AC \cdot BD = AD \cdot BC + AB \cdot CD$.

To show this using inversion in a circle, consider any circle $C(A, r)$ with center A and a large enough radius r so that the points B, C, D lie inside it. Let B', C', D' be the images of B, C, D, respectively, under inversion in $C(A, r)$. Then since B, C, D lie on a circle passing through the center A of $C(A, r)$, it follows that B', C', D' lie on a line.

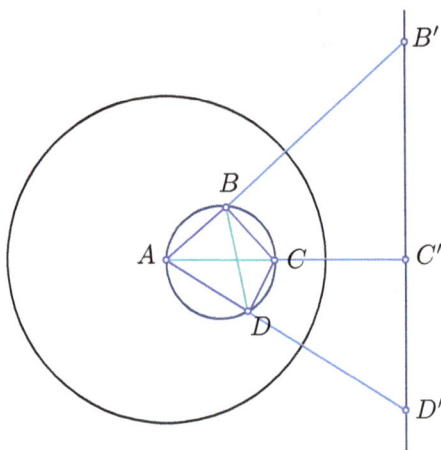

Thus $B'D' = B'C' + C'D'$. We will see that this delivers the conclusion in Ptolemy's Theorem!

To see this, let us first note that $\triangle ABC \sim \triangle AC'B'$, $\triangle ACD \sim \triangle AD'C'$ and $\triangle ABD \sim \triangle AD'B'$. These similarities give us, using the circle inversion relations (namely, $AB \cdot AB' = AC \cdot AC' = AD \cdot AD' = r^2$), that

$$\frac{B'C'}{BC} = \frac{AB'}{AC} = \frac{r^2}{AB \cdot AC},$$
$$\frac{C'D'}{CD} = \frac{AC'}{AD} = \frac{r^2}{AC \cdot AD},$$
$$\frac{B'D'}{BD} = \frac{AD'}{AB} = \frac{r^2}{AD \cdot AB}.$$

By using the expressions for $B'C', C'D'$ and $B'D'$ so obtained, in the equation $B'C' + C'D' = B'D'$, gives

$$\frac{r^2 \cdot BC}{AB \cdot AC} + \frac{r^2 \cdot CD}{AC \cdot AD} = \frac{r^2 \cdot BD}{AD \cdot AB}.$$

Finally, multiplying throughout by $AB \cdot AC \cdot AD/r^2$ gives the desired conclusion that $AD \cdot BC + AB \cdot CD = AC \cdot BD$.

Circular ↔ linear motion

Proposition 5.5 is the basis of the Peaucellier-Lipkin[18] linkage, which was the first[19] planar linkage that was able to transform circular motion into

[18] After (the French) Peaucellier (1832-1913) and (the Lithuanian) Lipkin (1846-1876).
[19] The interested reader is referred to the book [Bryant and Sangwin (2008)] for further details.

linear (that is, along a straight line) motion and vice versa. The linkage is depicted below.

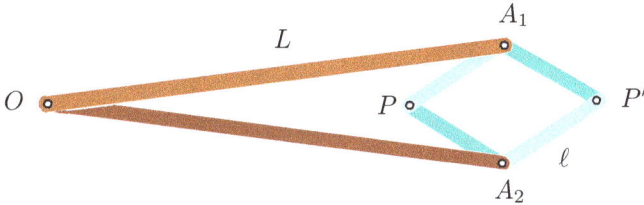

There are two arms, OA_1 and OA_2, of fixed lengths L, which are hinged at O (which is kept fixed at a point in the plane). There are also hinges at the points A_1, A_2, P, P', and four arms, $PA_1, PA_2, P'A_1$ and $P'A_2$, all having length ℓ, assumed to be strictly less than L.

Claim: P moves along a circle passing through O if and only if P' moves along a line.

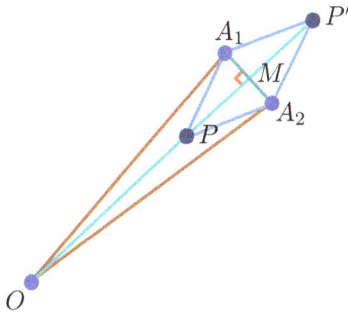

First we note that $A_1 P A_2 P'$ is a rhombus, and so its diagonals are the perpendicular bisectors of each other, and we call their intersection point M. So PP' is the perpendicular bisector of the segment $A_1 A_2$, and since $OA_1 = OA_2$, we conclude that O lies on PP', that is, O, P, P' are collinear. We have

$$
\begin{aligned}
OP \cdot OP' &= (OM - PM) \cdot (OM + MP') \\
&= (OM - PM) \cdot (OM + PM) = OM^2 - PM^2 \\
&= (OA_1^2 - A_1M^2) - (PA_1^2 - A_1M^2) \\
&= OA_1^2 - PA_1^2 = L^2 - \ell^2,
\end{aligned}
$$

where we have used Pythagoras's Theorem in the right angled triangles $\triangle OMA_1$ and $\triangle PMA_1$.

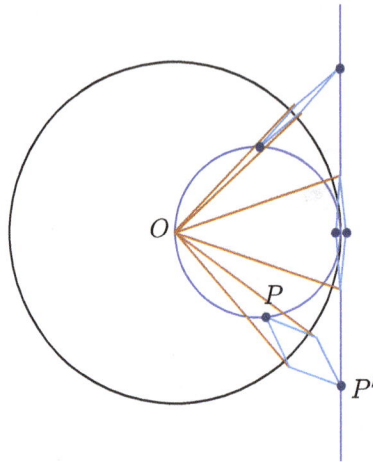

Now consider a circle $C(O, r)$ with center O and radius $r := \sqrt{L^2 - \ell^2}$ (> 0). As $OP \cdot OP' = r^2$, it follows that P, P' are images of each other under the inversion map in $C(O, r)$. By Proposition 5.5, we know that if we consider a circle $\mathcal{C}$ inside $C(O, r)$ that passes through O, then $\mathcal{C} \setminus \{O\}$ is mapped to a line, and vice versa. So it follows that as P moves along the circle $\mathcal{C}$ that passes through O and has diameter r, the image P' of P under the inversion map in the circle moves along a straight line, and vice versa. The picture above shows two "extremal" positions when the rhombus $A_1 P A_2 P'$ is collapsed one way or another, and an intermediate position of the linkage.

Notes

Exercise 5.3 is taken from [Trigg (1985)].
Exercise 5.9 was communicated by Victor Ufnarovski (Lund University).
Exercise 5.10 is based on [Kocik (2012)].
The proof of Theorem 5.17 is based on [Hess (2012)].
Exercise 5.32 is taken from [Gutenmacher and Vasilyev (1980)].
Exercise 5.37 is based on the article [Frohlinger and Hahn (2005)].
Exercise 5.41 is based on [Moreno and Garcia-Caballero (2013)].
The proof of Ptolemy's Theorem via inversion follows [Bogomolny (1997)].

Epilogue

Let us make a few concluding remarks here about what we have learnt, what has not been done, and where the reader might wish to go from here.

Euclid attempted to give an axiomatic formulation of plane geometry around 300 BC in his works *Elements* including concepts of points, straight lines and circles. We have loosely followed this trodden path. Later on, this axiomatic geometry was substantially refined, notably by David Hilbert, around the end of the 19th century in his book *Grundlagen der Geometrie* (The Foundations of Geometry). The goal in Hilbert's book was to start with a few axioms and derive all other propositions from them by purely logical deduction, without appealing to intuition. (We have broken this rule in our naïve treatment, since we have relied on our intuitive notions of points, lines, intersection, equality, coincidence, etc.) In the Hilbertian world, even the fundamental concepts are defined axiomatically, that is, with no content. This reads, for example, as follows. The elements P of a set $\mathbb{P}$ are called *points*. Certain subsets ℓ of $\mathbb{P}$ are called *straight lines*. Next, definitional statements and axioms are made on the interrelations between points and straight lines, between straight lines and straight lines, etc. Here are three examples:

(A1) If ℓ_1 and ℓ_2 are straight lines, then either $\ell_1 = \ell_2$ or $\ell_1 \cap \ell_2 = \emptyset$, or there exists exactly one point P such that $P \in \ell_1 \cap \ell_2$.

(A2) If P_1 and P_2 are two different points, $P_1 \neq P_2$, then there exists exactly one straight line ℓ such that $P_1 \in \ell$ and $P_2 \in \ell$.

(A3) (Euclid's Parallel Axiom). If ℓ is a straight line and P is a point such that $P \notin \ell$, then there exists exactly one straight line $L_{P,\ell}$ such that $P \in L_{P,\ell}$ and $\ell \cap L_{P,\ell} = \emptyset$.

This is of course not an exhaustive list, but a few examples to give the reader a flavour of the kind of things, and the type of language used there.

We also remark that Euclid's Parallel Axiom played a decisive role in the history and development of geometry. There were attempts to derive it from Euclid's other axioms. However, it was only in the first half of the 19th century that Gauss, Bolyai and Lobachevsky found out that Euclid's Parallel Axiom was not derivable from the other axioms of Euclid, since it can be replaced by the following "parallel axiom", while retaining the other axioms of Euclid, and yielding a consistent geometry (called *hyperbolic geometry*):

(A3′) If ℓ is a straight line, and P is a point not in ℓ, then there exist infinitely many straight lines ℓ' such that $P \in \ell'$ and $\ell \cap \ell' = \emptyset$.

On the other hand, this fits in the *Erlangen Programme*[1] of Felix Klein, which was initiated by him in the 1870s, where he had a "group-theoretic" viewpoint of Geometry. In this programme, Geometry was the study of a group of admissible symmetry transformations on a set, and their invariants (="geometric properties"). For example, in our Euclidean plane geometry, the set is the plane, the symmetry transformations in the plane are the congruent ones, that is, translations, rotations, and reflections, with distances and angles being geometric invariants.

This viewpoint of Klein also had a bearing on Physics. The Galilean relativity transformations in planar spacetime (that is, one space direction and one time direction) are the ones from Euclidean geometry. But, Einstein showed in his theory of special relativity, that the Galilean relativity transformations are not the ones describing the physics in inertial frames. Klein's viewpoint paved the way for Minkowski, who in 1907, showed that the theory of special relativity could be cast into a purely geometrical theory of space and time with an invariant based upon a variant (note the minus sign!) of Pythagoras's Theorem: $d\tau^2 = dt^2 - ds^2/c^2$, that is, the square of the "distance" $d\tau$ between two neighboring events in planar spacetime is defined as the difference between the square of the time separation dt and the square of the length separation ds between the two events measured in some inertial frame. This $d\tau$ is invariant under "Lorentz transformations". Here c is the velocity of light in vacuum. Subsequently in 1910 Klein showed how the Minkowski geometry of spacetime fits into the scheme of the Erlangen Programme. About the translation scheme between the relativity talk of physicists and his preferred geometric language, Klein said:

---

[1] Named after the University Erlangen-Nürnberg, where Klein worked.

If one wants to make a point of it, it would be all right to replace the phrase 'theory of invariants relative to a group of transformations' with the words 'relativity theory with respect to a group'.

We refer the interested reader to the book [Yaglom (1979)] for delving further into this subject.

Hints

Hints to the exercises from Chapter 1

Hint to Exercise 1.13

Each vertex can be joined to $n - 1$ other vertices, but two of them give sides, and so for getting a diagonal for each vertex there are $n - 1 - 2 = n - 3$ possible choices. Also one must take into account repetition while counting.

Hint to Exercise 1.14

See the picture below. There are infinitely many inscribed equilateral triangles in any given triangle.

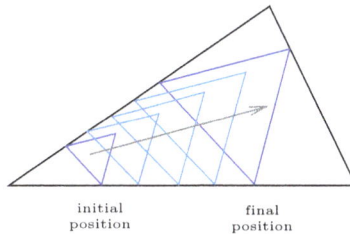

initial
position

final
position

Hint to Exercise 1.20

Consider all diagonals from one arbitrarily chosen vertex, "triangulating" the polygon.

Hint to Exercise 1.22

Do you see two superimposed triangles making the hexagram?

Hint to Exercise 1.23

Join the corner containing angle a to the one containing the angle c, and also join the corner containing angle a to the one containing the angle d.

Hint to Exercise 1.27

Focus on one little triangle shaded below. Do you see all the corner angles of the pentagram appearing in this little triangle?

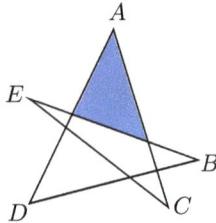

Hint to Exercise 1.28

Assume the contrary, and get a contradiction to the Exterior Angle Theorem.

Hint to Exercise 1.30

Note that Construction 1.1 works with a collapsible compass. If the given line segment is AB, and the given point is P, then think of an equilateral triangle of side length $PA + AB$ with a vertex P and a side lying on the ray $\overrightarrow{PA}$.

Hints to the exercises from Chapter 2

Hint to Exercise 2.3

To show that $AP + BP + CP < 2s$, proceed as follows. Extend BP to meet CA in E, and using the triangle inequalities in $\triangle ABE$ and $\triangle PCE$, show that $BP + PC < AB + AC$. Use similar inequalities with other combinations of the sides to obtain the desired result.

Hint to Exercise 2.4

Consider the reflection A' of A in the line ℓ.

Hint to Exercise 2.5

In any triangle $\triangle ABC$, think of a line parallel to the side BC, cutting AB and AC at points B', C'.

Hint to Exercise 2.6

What is the measure of each angle of a regular hexagon? See Exercise 1.20.

Hint to Exercise 2.7

There are several ways of going about this. Essentially, we are done if we manage to construct $1°$. And for this, is the following of any use?

$$1 = 20 - 19 = 2 \cdot 10 - 19 = 2 \cdot (190 - 180) - 19 = 2 \cdot (10 \cdot 19 - 180) - 19.$$

Hint to Exercise 2.8

Think of an "equilateral triangular lattice of points".

Hint to Exercise 2.9

Show that the minimizing position of the line is such that P is the midpoint of the line segment passing through P (with endpoints on the two bounding rays of the given angle).

Hint to Exercise 2.10

Start with a scalene triangle, and construct the angle bisector of any angle and the perpendicular bisector of the opposite side. What do you notice about their point of intersection?

Hint to Exercise 2.11

There are several ways of showing this, and the idea is that starting with two arbitrary points A, B, we need to produce triangles which share an incircle so as to get useful relations, leading to the relation $f(A) = f(B)$. One way to proceed is to construct the regular pentagon with AB as one side, and produce AE, CD to meet at a point, say P. Note that the triangles $\triangle ADP, \triangle CEP$ share their incircle (why?), and this gives $f(A) + f(D) = f(C) + f(E)$. Using other similar relations for the values of f on the vertices of the pentagon, complete the proof.

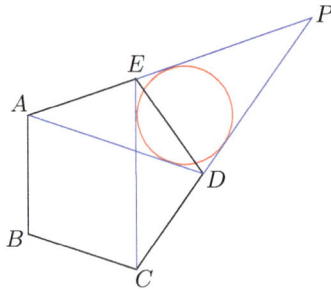

Hints to the exercises from Chapter 3

Hint to Exercise 3.2

If the diagonals AC and BD intersect at O, then show that $\triangle AOD \simeq \triangle COB$.

Hint to Exercise 3.3

Reflect a vertex A of the triangle $\triangle ABC$ through the midpoint D of the opposite side BC in order to obtain a point A'. Consider the parallelogram $ABA'C$. Use the triangle inequality.

Hint to Exercise 3.4

If the centers of the three circles are C_1, C_2, C_3, their common intersection point is O, and the other intersection points are A_{12}, A_{23}, A_{31}, then show that $\triangle C_1 C_2 C_3 \simeq \triangle A_{23} A_{31} A_{12}$. Note that the former triangle has circumradius equal to the radius of each circle!

Hint to Exercise 3.5

Consider the center of a square.

Hint to Exercise 3.6

Draw a diagonal in the given quadrilateral.

Hint to Exercise 3.7

Midpoint Theorem galore!

Hint to Exercise 3.8

Consider the quadrilateral $BQDS$. Use Midpoint Theorem 2.

Hint to Exercise 3.9

Draw a diagonal in the trapezium, and consider its midpoint.

Hint to Exercise 3.10

If $ABCD$ is the trapezium, and P, Q are the midpoints of the diagonals AC, BD, respectively, then join P to the midpoint M of the side BC.

Hint to Exercise 3.11

The procedure is: join the base point D of an altitude AD lying inside a triangle $\triangle ABC$ to the midpoints F, E of the other two sides AB, AC. (In order to see the justification of this procedure, join the midpoints F, E.)

Hint to Exercise 3.12

Show that the triangles obtained in consecutive steps share their centroid.

Hint to Exercise 3.13

Draw the angle bisector of $\angle A$, and let it meet BC in D'.

Hint to Exercise 3.14

Draw a diagonal.

Hint to Exercise 3.15

Consider the expressions for the areas in terms of the three different sides and the corresponding altitudes to the sides.

Hint to Exercise 3.16

Join the incenter to the three vertices, and add up the areas of the three little triangles so obtained, to get the area of the triangle.

Hint to Exercise 3.17

Use the previous exercise.

Hint to Exercise 3.19

Show that in $\triangle A'B'C'$, the length of the altitude to the base formed by the reflection of the vertices forming the hypotenuse in $\triangle ABC$ is thrice the length of the altitude to the hypotenuse in $\triangle ABC$.

Hint to Exercise 3.20

The area of the rectangle is twice the area of the staircase region.

Hint to Exercise 3.21

Join the interior point to the three vertices, forming smaller triangles, and express the area of the equilateral triangle in terms of these smaller triangle areas.

Hint to Exercise 3.22

Join BB'. Compare the area of $\triangle AC'B'$ with that of $\triangle ABB'$, and the area of $\triangle ABB'$ in turn with that of $\triangle ABC$.

Hint to Exercise 3.25

Double application of the Pythagoras Theorem.

Hint to Exercise 3.26

Imagine the cup made of paper, cutting free the base, and then making a vertical cut along the resulting cylinder. What do we obtain?

Hint to Exercise 3.28

For part (3), what does $c^2 = a^2 + b^2$ say in terms of the areas shown in the picture?

Hint to Exercise 3.29

No.

Hint to Exercise 3.34

Let x be the distance travelled by the snake (and the peacock). Use Pythagoras's Theorem to set up an algebraic equation in x.

Hint to Exercise 3.36

Consider another triangle $\triangle A'B'C'$ with $A'B' = AB$, $A'C' = AC$ and $\angle A' = 90°$. Are $\triangle ABC$ and $\triangle A'B'C'$ congruent?

Hint to Exercise 3.37

Use Apollonius's Theorem.

Hint to Exercise 3.38

Use Pythagoras's Theorem and Heron's Formula.

Hints to the exercises from Chapter 4

Hint to Exercise 4.3

Draw a line through P parallel to AB, meeting BC in, say, Q.

Hint to Exercise 4.4

If the given lines intersect at A, then try to locate points B, C on the two given lines such that P is the orthocenter of $\triangle ABC$.

Hint to Exercise 4.5

Use Ceva's Theorem and the Basic Proportionality Theorem.

Hint to Exercise 4.8

Draw the external angle bisector at A, and let it meet BC extended in D'.

Hint to Exercise 4.9

(1): Considering the triangle pairs $(\triangle CML, \triangle CAD)$ and $(\triangle CMN, \triangle CAB)$, show that $CM/AC = MN/AB = ML/AD$.
Also show that $MP/PB = MN/AB = ML/BD$ by considering the two triangle pairs $(\triangle MNP, \triangle BAP)$ and $(\triangle MLP, \triangle BDP)$.
(2): Extend one of the sides of the square and take any point on this line. Join this point to one of the other two vertices, and use the previous part!

Hint to Exercise 4.10

If the diagonals intersect at P, note that $\triangle APB \sim \triangle CPD$, and use Pythagoras's Theorem in each of the four triangles $\triangle APB$, $\triangle BPC$, $\triangle CPD$, $\triangle DPA$.

Hint to Exercise 4.12

Consider the point F' which is the intersection point of the extension of DE and AB.

Hint to Exercise 4.13

Join B to D.

Hint to Exercise 4.14

Apply Menelaus's Theorem to $\triangle ABA'$ with the line CC' cutting it, and determine the ratio $XA' : AA'$.

Hint to Exercise 4.15

For (1), use the similarity of $\triangle APM$ and $\triangle AA'P$. For part (2), use part (1). For part (3), think of the case when $AP = PM$ as before, but $AA' = nAB$, and show that AM is then AB/n.

Hint to Exercise 4.16

Show that $\triangle ABO \sim \triangle OB'A'$ to deduce that $\angle AOA' = 90°$, and then use Pythagoras's Theorem.

Hint to Exercise 4.17

(If the point is at one of the vertices, then two of the areas collapse to 0, while the third area becomes the full area of the triangles, and so the result is true in this case.) Now consider the next extreme case, where the point lies on one of the sides of the triangle. In this case, only one of the three areas A_1, A_2, A_3, say A_3 becomes 0. In this case, both of the little triangles are similar to the big triangle. Can you show that in this case we have $\sqrt{A} = \sqrt{A_1} + \sqrt{A_2}$? Finally use the previous case to derive the result when the point P is in "general position" inside the triangle.

Hint to Exercise 4.18

The formula
$$r = \frac{2\Delta}{a+b+c}$$
for the inradius of a triangle $\triangle ABC$ with area Δ and side lengths a, b, c (see Exercise 3.16) shows that the inradius scales in proportion to the sides. Now use the fact that the two little triangles are similar to the big triangle.

Hint to Exercise 4.19

Using the notation from the proof of Theorem 4.10, we know $ax \geq cm + bn$. Use the Arithmetic Mean-Geometric Mean Inequality (see Exercise 5.7) to underestimate the right-hand side. Multiply all such inequalities!

Hints to the exercises from Chapter 5

Hint to Exercise 5.4

Use Pythagoras's Theorem.

Hint to Exercise 5.5

Complete the hexagon with center A and vertices B, C by successive reflection.

Hint to Exercise 5.6

(1),(2): Yes using the Internal/External Angle Bisector Theorems. (3): $90°$.

Hint to Exercise 5.7

(2): Use the similar triangles $\triangle ABB'$ and $\triangle B'BA'$. For (3), find OB and the use Pythagoras's Theorem in $\triangle O'OB$. For (4), note that OB' is the radius of the semicircle, and then use the similarity of the triangles $\triangle OBB'$ and $\triangle BCB'$.

Hint to Exercise 5.8

Use similarity.

Hint to Exercise 5.9

Show that the locus is the interior of a circle with diameter BC, except the diameter BC itself.

Hint to Exercise 5.11

Consider the line passing through A and B, and the points on the circle on either side of this line. Compare the angles subtended at these points with the angles subtended at the points of contact with the two smaller circles.

Hint to Exercise 5.14

Assuming that τ is rational, consider the smallest similar golden rectangle with integer sides to arrive at a contradiction.

Hint to Exercise 5.15

(1): Divide the recurrence rule throughout by F_n.

Hint to Exercise 5.16

Complete the right triangle to a rectangle, and consider its circumscribing circle.

Hint to Exercise 5.17

Use Ptolemy's Theorem for the cyclic quadrilateral with vertices A, B, C, P.

Hint to Exercise 5.18

Use Theorem 5.14.

Hint to Exercise 5.19

Use the previous theorem.

Hint to Exercise 5.20

If we choose b as the base, then express the altitude to this base in terms of the other sides a, c and the diameter $2R$ by using similar triangles and the fact that the angle subtended by the diameter passing through A at the point B is $90°$.

Hint to Exercise 5.21

Express Δ in terms of a, b, c and x, y, z. Use the Arithmetic Mean-Geometric Mean Inequality, saying that if r_1, r_2, r_3 are three nonnegative real numbers, then $(r_1 + r_2 + r_3)/3 \geq \sqrt[3]{r_1 r_2 r_3}$, and equality holds if and only if $r_1 = r_2 = r_3$. Also use Exercise 5.20.

Hint to Exercise 5.22

Use Theorem 5.14.

Hint to Exercise 5.24

Try guessing the answer by looking at what happens when a pair of parallel lines intersect a circle.

Hint to Exercise 5.25

Extend the nonparallel sides.

Hint to Exercise 5.26

Show that opposite angles are supplementary in $A'B'C'D'$.

Hint to Exercise 5.27

(1) For cyclicity, note that the opposite angles are supplementary. Then use $\angle X'Y'D = \angle DAX' = \angle XAP = \angle XYP$, and $\angle XPY = 180° - \angle A = \angle Y'DX'$ to conclude the similarity of the triangles.
(2) Consider the areas of $\triangle ABD''$ and $\triangle AD''C$, and think of moving P down to coincide with D''.
(3) Ceva.
(4) Let the distances of the point P to the sides AB, AC, BC, be denoted by x, y, z. Note that the area of the triangle $\triangle ABC$ is $(cx + by + az)/2$, and use Cauchy-Schwarz.

Hint to Exercise 5.30

Let O'' be a point such that D is the midpoint of OO''. Show that O' and O'' coincide by observing that $OBO''C$ has to be a parallelogram, and showing that $\angle ACO'' = 90°$.

Hint to Exercise 5.31

Ceva's Theorem.

Hint to Exercise 5.32

The answer is the diameter of the big circle passing through the current position of the point. Use the arc length formula and Theorem 5.6.

Hint to Exercise 5.33

Use Theorem 5.22.

Hint to Exercise 5.34

Use Theorem 5.21.

Hint to Exercise 5.35

Use Theorem 5.21.

Hint to Exercise 5.37

Consider the family of circles which has AB as a chord. Among these, take the one which is tangential to the floor of the cinema, and show that the point P maximizing the viewing angle is the point of tangency.

Hint to Exercise 5.38

Express the side lengths in terms of the lengths of the tangents drawn to the incircle from the vertices of the right triangle.

Hint to Exercise 5.40

For the case when AB is not parallel to ℓ, we would be done if we determine the point of tangency T (since then we can just construct the circumcircle of $\triangle ABT$). But if P denotes the point of intersection of ℓ with AB (extended), then note that $PT = \sqrt{PA \cdot PB}$, the geometric mean of PA and PB. Recall Exercise 5.7.

Hint to Exercise 5.41

Use Pythagoras's Theorem.

Solutions

Solutions to the exercises from Chapter 1

Solution to Exercise 1.1

Let the angle have measure $\alpha°$. Then its complement has measure $90° - \alpha°$. It is given that $\alpha = 20 + (90 - \alpha)$, so that $\alpha = 55$. So the measure of the angle is $55°$.

Solution to Exercise 1.2

Let the angle have measure $\alpha°$. Then its supplement has measure $180° - \alpha°$. It is given that $\alpha = 3(180 - \alpha)$, so that $4\alpha = 3 \cdot 180$. So $\alpha = 135$.

Solution to Exercise 1.3

We have $2x + 3x = 180$, and so $x = 180/5 = 36$.

Solution to Exercise 1.4

Extend $\overrightarrow{OA}$ to the opposite side to obtain the straight line $\overleftrightarrow{A'A}$.

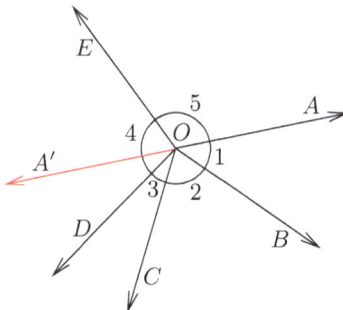

Then $\angle 1 + \angle 2 + \angle 3 + \angle DOA' = 180°$, and $\angle 5 + \angle EOA' = 180°$. We also have $\angle EOA' + \angle DOA' = \angle 4$, and so it follows by adding the above two equations that

$$\angle 1 + \angle 2 + \angle 3 + \angle 4 + \angle 5 = \angle 1 + \angle 2 + \angle 3 + \angle DOA' + \angle EOA' + \angle 5 = 180° + 180° = 360°.$$

Solution to Exercise 1.5

We have $\angle COD + \angle EOC = \dfrac{\angle COA}{2} + \dfrac{\angle BOC}{2} = \dfrac{180°}{2} = 90°$. So $\overrightarrow{OE} \perp \overrightarrow{OD}$.

Solution to Exercise 1.6

As opposite angles are equal, we obtain $5y + 2y + 5y = 180$, giving $12y = 180$. Hence $y = 15$.

Solution to Exercise 1.7

$\angle COB + \angle BOD = 180°$. But $2\angle BOP = \angle BOD = \angle COA = 2\angle COQ$. Thus

$$\begin{aligned}180° &= \angle COB + \angle BOD = \angle COB + 2\angle COQ \\ &= \angle COB + \angle COQ + \angle COQ = \angle COB + \angle COQ + \angle BOP \\ &= \angle QOB + \angle BOP.\end{aligned}$$

Hence $\angle QOB$ and $\angle BOP$ are supplementary. So P, O, Q are collinear.

Solution to Exercise 1.8

Let ℓ, m intersect in the point P. By the Parallel Postulate, there exists a unique line $L_{P,n}$ which passes through P and is parallel to n. As m passes through P and is parallel to n, it follows that $L_{P,n} = m$. Since $\ell \neq m$, and as ℓ passes through P, we conclude using the uniqueness of $L_{P,n}$ that $\ell \not\parallel n$. Consequently ℓ intersects n.

Solution to Exercise 1.9

Suppose that C does not lie on $\overleftrightarrow{AB}$. Then the lines $\overleftrightarrow{AB}$ and $\overleftrightarrow{AC}$ are both parallel to ℓ, and pass through A, contradicting the Parallel Postulate. Thus C lies on $\overleftrightarrow{AB}$, that is, the points A, B, C are collinear.

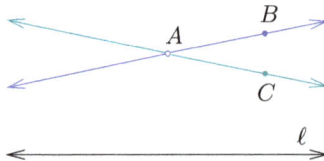

Solution to Exercise 1.10

We check that $\parallel$ is reflexive, symmetric and transitive.

(ER1) (Reflexivity) By definition, a line is considered to be parallel to itself, and so for every line ℓ in the plane, $\ell \parallel \ell$.

(ER2) (Symmetry) If $\ell \parallel \ell'$, then either they are coincident or don't intersect. In either case, we have $\ell' \parallel \ell$.

(ER3) (Transitivity) Suppose that $\ell \parallel \ell'$ and $\ell' \parallel \ell''$.
If $\ell = \ell'$, then it follows from $\ell' \parallel \ell''$ that $\ell \parallel \ell''$, and we are done.
If $\ell' = \ell''$, then it follows from $\ell \parallel \ell'$ that $\ell \parallel \ell''$, and we are done.
Now suppose that $\ell \neq \ell'$ and $\ell'' \neq \ell'$. Let us assume that $\ell \neq \ell''$ and that ℓ and ℓ'' intersect in a point P. This P lies outside ℓ'. Then through the point P outside ℓ', there are two distinct lines, namely ℓ and ℓ'', which are both parallel to ℓ', a contradiction to the Parallel Postulate. This contradiction shows that our original assumption must be false, that is, either $\ell = \ell''$ or it must be the case that ℓ, ℓ'' don't intersect. In either case, $\ell \parallel \ell''$.

Solution to Exercise 1.11

Let us draw the normals BM and CN to the two mirrors at the points B, C, respectively, as shown. Considering the transversal BM intersecting the two mirrors, we know that the consecutive interior angles $\angle CMB$ and $\angle MBN$ are supplementary. As $\angle MBN = 90°$, it follows that $\angle CMB = 90°$. Now consider the transversal CM to CN and MB. We have $\angle NCM = \angle CMB = 90°$, and so they are supplementary. But they form a pair of consecutive interior angles for the transversal CM to the two lines CN and MB, and so we conclude that $CN \parallel MB$.

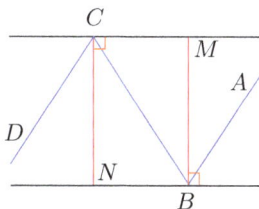

By the Law of Reflection, $\angle ABM = \angle MBC$. As $CN \parallel MB$, the alternate angles to the transversal CB are equal, and so $\angle NCB = \angle CBM$. Finally, again by the Law of Reflection, $\angle DCN = \angle NCB$. We have

$$\angle DCB = \angle DCN + \angle NCB = 2\angle NCB$$
$$= 2\angle CBM = \angle CBM + \angle MBA = \angle CBA,$$

and so if we consider the transversal CB to the two lines CD and AB, we have obtained that the alternate interior angles $\angle DCB = \angle CBA$. Thus $AB \parallel CD$.

Solution to Exercise 1.12

Let us draw the normals to the two mirrors at the points B, C, and suppose that they meet at the point P. Let the "corner" of the two mirrors be the point Q.

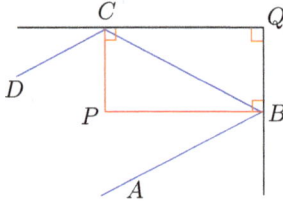

As $\angle CQB = 90° = \angle PCQ$, these two angles are supplementary. But they are the consecutive interior angles for the pair of lines PC, BQ with the transversal CQ. Hence $CP \parallel BQ$, and so the alternate interior angles with the transversal BC are equal, and in particular, $\angle PCB = \angle CBQ$. By the Law of Reflection, $\angle ABP = \angle PBC$ and $\angle BCP = \angle PCD$. Thus we have

$$\begin{aligned}
\angle ABC &= \angle ABP + \angle PBC = 2\angle PBC = 2(90° - \angle CBQ) \\
&= 180° - 2\angle CBQ = 180° - 2\angle PCB \\
&= 180° - (\angle PCB + \angle PCD) \\
&= 180° - \angle BCD.
\end{aligned}$$

So the angles $\angle ABC$ and $\angle BCD$ are supplementary. But they are consecutive interior angles for the pair of lines AB, CD with the transversal BC. Hence $AB \parallel CD$.

Solution to Exercise 1.13

Each vertex can be joined to $n-1$ other vertices, but two of them give sides, and so for getting a diagonal for each vertex there are $n-1-2 = n-3$ possible choices. Hence the number of pairs of vertices which describe a diagonal is $n \cdot (n-3)$. But $P_i P_j$ and $P_j P_i$ describe the same line segment, and so we have counted twice the total number of diagonals by counting the total number of pairs of vertices giving a diagonal. Hence the number of diagonals in a regular polygon with n sides is

$$\frac{n \cdot (n-3)}{2}.$$

When $n = 5$, we have $\dfrac{n \cdot (n-3)}{2} = \dfrac{5 \cdot 2}{2} = 5$, as expected.

Solution to Exercise 1.14

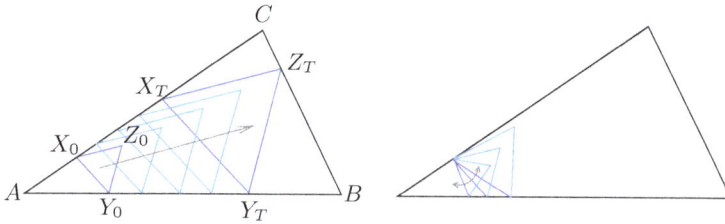

We start with an arbitrarily aligned tiny line segment X_0Y_0 whose vertices lie on the sides AB, AC of the original triangle $\triangle ABC$ near the vertex A of the original triangle. Then we complete this to an equilateral triangle with the third vertex Z_0 lying away from the chosen vertex A. Now we consider the two movable points X_t on AC and Y_t on AB, such that when $t = 0$, X_t coincides with X_0 and Y_t with Y_0. Moreover, the points move so that X_tY_t stays parallel to the original segment X_0Y_0, while scaling up the equilateral triangle $\triangle X_tY_tZ_t$. The third vertex Z_t of the equilateral triangle stays inside the given triangle, but at a certain point of time $t = T$, it will touch the side BC. At this moment, we have got an equilateral triangle $\triangle X_TY_TZ_T$, all of whose vertices lie on the sides of the original triangle.

Since the original alignment of X_0Y_0 was arbitrary (see the right picture above), we see that there are infinitely many such inscribed equilateral triangles in the given triangle.

Solution to Exercise 1.15

The interior of an angle is convex, but its exterior is nonconvex. The following pictures give the rationale behind these answers.

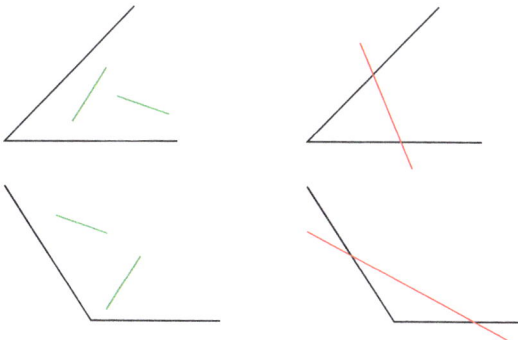

Interior is convex   Exterior is not convex

Solution to Exercise 1.16

We have $16, 7, 3, 1$ triangles with side lengths $1, 2, 3, 4$ units, respectively, and so the total number of triangles is $1 + 3 + 7 + 16 = 27$.

Solution to Exercise 1.17

Let the angles be $x, 2x, 3x$ degrees. Since the sum of the angles of a triangle is $180°$, we have $x + 2x + 3x = 180$, that is, $6x = 180$, and so $x = 30$. So the angles of the triangle are $30°, 60°, 90°$.

Solution to Exercise 1.18

(1) No, since otherwise the sum of the angles would be more than $180°$.

(2) No, since otherwise the sum of the angles would be more than $180°$.

(3) Yes. Consider an equilateral triangle, in which all angles are equal to $60°$. (We will prove this later.)

(4) No, since otherwise the sum of the angles would be more than $180°$.

(5) No, since otherwise the sum of the angles would be less than $180°$.

(6) Yes. An equilateral triangle has all angles equal to $60°$.

Solution to Exercise 1.19

If the two angles have measures x, y degrees, then $x + y = 90$ and $x - y = 30$. Adding, we obtain $2x = 120$, and so $x = 60$. This gives $y = 90 - x = 90 - 60 = 30$. The third angle z can be found using the fact that the sum of the three angles in a triangle is $180°$: $z = 180 - (30 + 60) = 180 - 90 = 90$. Thus the angles of the triangle are $30°, 60°, 90°$.

Solution to Exercise 1.20

Choose any vertex of the polygon, and consider the $n - 3$ diagonals from that vertex. This divides the polygon into $n - 3 + 1 = n - 2$ triangles. The sum of the angles of the polygon will be the sum of all the angles of these little triangles. Thus the sum of all the angles of the polygon is $(n - 2) \cdot 180°$.

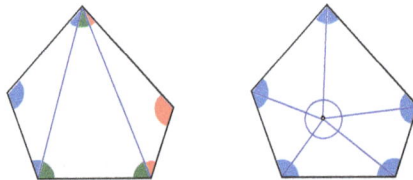

(Alternatively, we may consider any interior point O of the polygon which is joined to each of the vertices, forming n triangles. Now the sum of the angles of those angles of these small triangles which have the vertex at O is $360°$, and so the sum of the left over angles of these little triangles is $n \cdot 180° - 360° = (n-2) \cdot 180°$. But the sum of these is clearly the sum of the angles of the polygon.)

When $n = 4$, we have $(n-2) \cdot 180° = (4-2) \cdot 180° = 2 \cdot 180° = 360°$, as expected.

In a regular polygon, all angles are equal, and so the measure of each angle is $\dfrac{(n-2) \cdot 180}{n}°$.

We have $\dfrac{(n-2) \cdot 180}{n} = 120$, and so $n = 6$, that is, the polygon is a hexagon.

Solution to Exercise 1.21

By Exercise 1.20, each angle in a regular pentagon is $\dfrac{(5-2) \cdot 180}{5}° = 108°$.

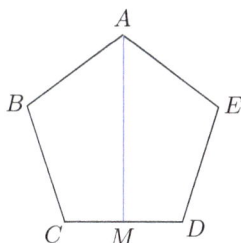

Thus $\angle AMC = 360° - \left(\angle B + \angle C + \dfrac{\angle A}{2} \right) = 360° - (108° + 108° + 54°) = 90°$.

Solution to Exercise 1.22

Considering $\triangle AEC$, we have $\angle A + \angle C + \angle E = 180°$, and similarly by looking at $\triangle BDF$ we obtain $\angle B + \angle D + \angle F = 180°$.

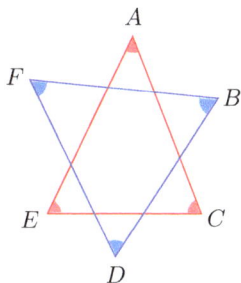

Adding these, we obtain $\angle A + \angle B + \angle C + \angle D + \angle E + \angle F = 180° + 180° = 360°$.

Solution to Exercise 1.23

From the picture, $a+b+c+d = 180°+180°+\alpha+\beta$, and since $\ell \parallel \ell'$, $\alpha+\beta = 180°$ (being consecutive interior angles). Thus $a + b + c + d = 3 \cdot 180° = 540°$.

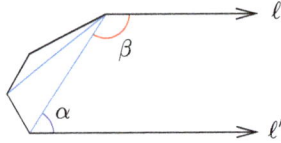

Solution to Exercise 1.24

We have $\angle OBC = \dfrac{\angle B}{2}$ and $\angle OCB = \dfrac{\angle C}{2}$.

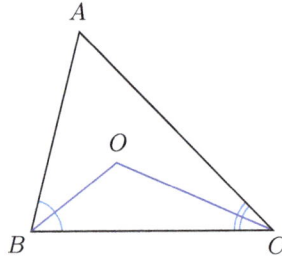

Thus $\angle OBC + \angle OCB = \dfrac{\angle B + \angle C}{2} = \dfrac{180° - \angle A}{2} = 90° - \dfrac{\angle A}{2}$.

Hence $\angle BOC = 180° - (\angle OBC + \angle OCB) = 180° - \left(90° - \dfrac{\angle A}{2}\right) = 90° + \dfrac{\angle A}{2}$.

Solution to Exercise 1.25

We have

$$\angle OBC = \frac{\angle B'BC}{2} = \frac{\angle A + \angle C}{2}, \text{ and } \angle OCB = \frac{\angle C'CB}{2} = \frac{\angle A + \angle B}{2}.$$

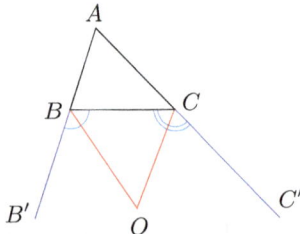

Adding these,

$$\angle OBC + \angle OCB = \angle A + \frac{\angle B + \angle C}{2} = \angle A + \frac{180° - \angle A}{2} = \frac{\angle A}{2} + 90°.$$

So $\angle BOC = 180° - (\angle OBC + \angle OCB) = 180° - \left(\frac{\angle A}{2} + 90°\right) = 90° - \frac{\angle A}{2}.$

Solution to Exercise 1.26

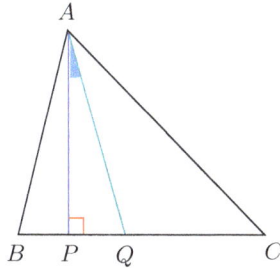

We have

$$\angle PAQ = 90° - \angle AQB = 90° - (\angle QAC + \angle C)$$

$$= 90° - \frac{\angle A}{2} - \angle C = 90° - \frac{180° - \angle B - \angle C}{2} - \angle C = \frac{\angle B - \angle C}{2}.$$

Solution to Exercise 1.27

Let us look at the picture on the right below. The top angle is $\angle A$, while we have two base angles. Each of these base angles is an exterior angle, as shown in the picture on the left and the middle picture. In the leftmost picture we see that the exterior angle is $\angle C + \angle E$, while in the middle picture we see that the exterior angle is $\angle B + \angle D$. Consequently, by looking at the angles in the little triangle in the picture on the right, we see that $\angle A + (\angle B + \angle D) + (\angle C + \angle E) = 180°$, that is, $\angle A + \angle B + \angle C + \angle D + \angle E = 180°$.

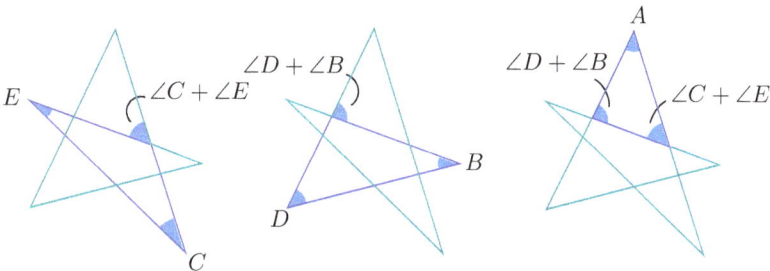

Solution to Exercise 1.28

Let the altitude dropped from the vertex A meet the line containing the opposite side BC at a point D outside AB. Without loss of generality, we may assume that the point D is to the left of B. (Otherwise we can swap the labels B and C.)

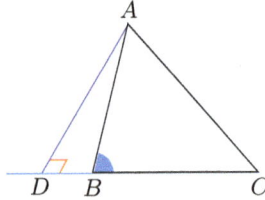

Now consider triangle $\triangle ABD$. $\angle ABC$ is an exterior angle to this triangle, and so it must equal the sum of the opposite interior angles, that is,

$$\angle ABC = \angle DAB + \angle ADB = \angle DAB + 90° > 90°,$$

which is a contradiction to the given fact that the triangle $\triangle ABC$ is an acute angled triangle.

Remark 6.1. Later on we will learn that the altitudes in any triangle are concurrent, and their meeting point is called the orthocenter. Based on this exercise, we can conclude that in an acute angled triangle, the orthocenter lies in the interior of the triangle. In fact one can also prove the converse. In a right angled triangle, the orthocenter lies at the vertex of the right angle, and in an obtuse angled triangle, the orthocenter lies outside the triangle.

Solution to Exercise 1.29

By the Corresponding Angles Axiom, it follows that the two lines are not parallel. So they must intersect at exactly one point. Suppose that on the contrary they meet at the side opposite to what is claimed. Suppose that the angle made by the intersecting lines is $\theta > 0$.

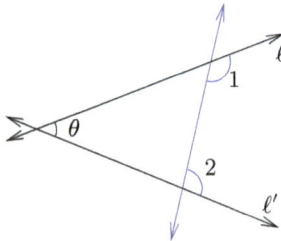

Then by referring to the picture above, we note that $\angle 1 = \theta + (180° - \angle 2)$, while it is given that $\angle 1 + \angle 2 < 180°$, and so we arrive a contradiction.

Solution to Exercise 1.30

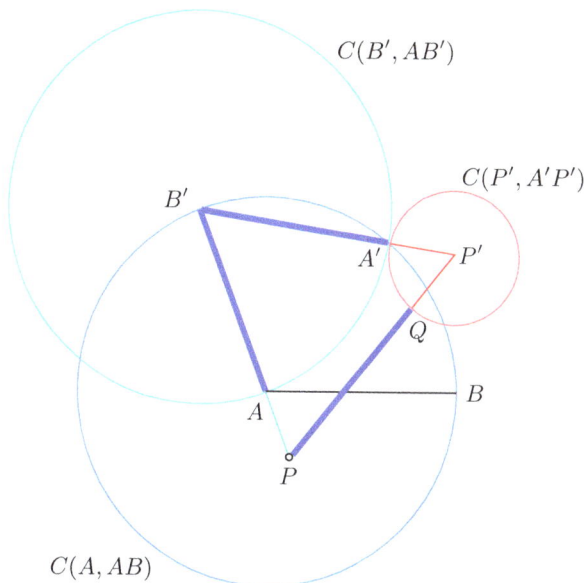

Step 1. Draw the ray $\overrightarrow{PA}$ starting from the point A.

Step 2. With A as center and radius AB, draw the circle $C(A, AB)$.
Let B' be the point of intersection of the circle $C(A, AB)$ and the ray $\overrightarrow{PA}$.

Step 3. Construct an equilateral triangle $\triangle B'PP'$ with base $B'P$ following the steps from Construction 1.1.

Step 4. With center as the point B' and radius AB', draw the circle $C(B', AB')$.
Let A' be the point of intersection of the circle $C(A, AB')$ and the segment $B'P'$.

Step 5. With P' as the center and radius $A'P'$, draw a circle of radius $A'P'$.
Let Q be the point of intersection of PP' and $C(P', A'P')$.

Then PQ is the required line segment with length equal to AB.

Justification: As B, B' lie on $C(A, AB')$, it follows that $AB = AB'$. As A, A' lie on $C(B', AB')$, we also have $AB' = A'B'$. Hence $AB = AB' = A'B'$. Moreover, as A', Q lie on $C(P', A'P')$, we obtain $A'P' = QP'$. As $\triangle PB'P'$ is equilateral, $B'P = B'P' = PP'$. So $PQ = PP' - QP' = B'P' - A'P' = B'A' = B'A = AB$.

Remark 6.2. In fact, the point A' is simply the intersection of $C(A, AB)$ and the line segment $B'A'$, and this can be shown once we know that the angles in an equilateral triangle are all equal to $60°$. However, this has not yet been shown, and so we preferred to use A' as the point of intersection of the circle $C(B', AB')$ and the line segment $B'P'$.

Solutions to the exercises from Chapter 2

Solution to Exercise 2.1

We check that the congruency relation is reflexive, symmetric and transitive:

(1) Reflexivity: A triangle $\triangle ABC$ is obviously congruent to itself, since the corresponding sides are equal and the corresponding angles are equal, that is,

$$AB = AB, \quad BC = BC, \quad CA = CA$$
$$\angle A = \angle A, \quad \angle B = \angle B, \quad \angle C = \angle C.$$

(2) Symmetry: If $\triangle ABC \simeq \triangle A'B'C'$, then

$$AB = A'B', \quad BC = B'C', \quad CA = C'A'$$
$$\angle A = \angle A', \quad \angle B = \angle B', \quad \angle C = \angle C'$$

and by reading these equalities backwards, we see that $\triangle A'B'C' \simeq \triangle ABC$.

(3) Transitivity: Let $\triangle ABC \simeq \triangle A'B'C'$ and $\triangle A'B'C' \simeq \triangle A''B''C''$. Then

$$AB = A'B', \quad BC = B'C', \quad CA = C'A'$$
$$\angle A = \angle A', \quad \angle B = \angle B', \quad \angle C = \angle C'$$

and

$$A'B' = A''B'', \quad B'C' = B''C'', \quad C'A' = C''A''$$
$$\angle A' = \angle A'', \quad \angle B' = \angle B'', \quad \angle C' = \angle C''.$$

Putting these together, we obtain also that

$$AB = A''B'', \quad BC = B''C'', \quad CA = C''A''$$
$$\angle A = \angle A'', \quad \angle B = \angle B'', \quad \angle C = \angle C'',$$

that is, $\triangle ABC \simeq \triangle A''B''C''$.

Solution to Exercise 2.2

By the triangle inequality, $b + c > a$, and so we have

$$s - a = \frac{a+b+c}{2} - a = \frac{b+c-a}{2} > 0.$$

Similarly, we can also show, using $c + a > b$ and $a + b > c$, that $s - b > 0$ and $s - c > 0$. Thus we have $s > a$, $s > b$ and $s > c$, and so $s > \max\{a, b, c\}$.

Solution to Exercise 2.3

By the Triangle Inequality in $\triangle ABE$,

$$AB + AE > BE = BP + PE. \tag{6.1}$$

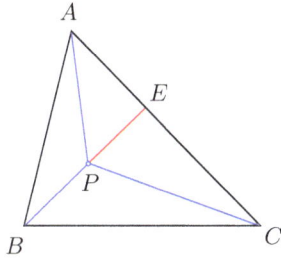

By the Triangle Inequality in $\triangle PCE$,

$$PE + EC > CP. \tag{6.2}$$

Adding (6.1) and (6.2), we obtain $AB + AE + EC + \cancel{PE} > BP + CP + \cancel{PE}$, that is,

$$AB + AE + EC = AB + AC > BP + CP. \tag{6.3}$$

Similarly, one can show that also

$$AC + BC > AP + BP, \tag{6.4}$$
$$BC + AB > CP + AP. \tag{6.5}$$

Adding (6.3), (6.4) and (6.5), $4s = 2(AB + BC + CA) > 2(AP + BP + CP)$, that is, $AP + BP + CP < 2s$. So we've got the second inequality asked in the exercise.

Now we will show the first inequality. By the Triangle Inequalities in $\triangle ABP$, $\triangle BCP$, $\triangle CAP$,

$$AB < AP + BP,$$
$$BC < BP + CP,$$
$$CA < CP + AP.$$

Adding these, we obtain $2s = AB + BC + CA < 2(AP + BP + CP)$, that is,

$$s < AP + BP + CP.$$

Solution to Exercise 2.4

Consider the reflection A' of the point A in the line ℓ, namely the point A' such that AA' is perpendicular to ℓ, and moreover, if AA' intersects ℓ in X, then also $AX = A'X$. See the picture below.

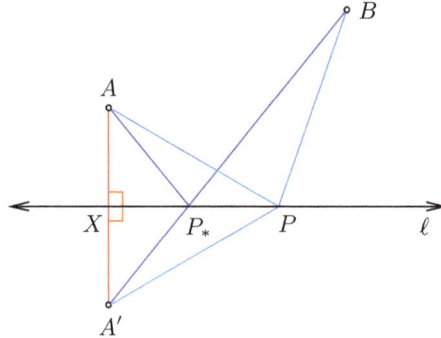

Join A to P, P to A', P to B, and A' to B. Let $A'B$ meet ℓ in the point P_*. Also join A to P_*. By the SAS Congruency Rule, $\triangle APX \simeq \triangle A'PX$ (as we have $AX = A'X$, the side PX is common to both, and the included angles $\angle AXP$ and $\angle A'XP$ are both $90°$). So $AP = A'P$ (CPCT). Similarly, $AP_* = A'P_*$ since $\triangle AP_*X \simeq \triangle A'P_*X$. Hence by the triangle inequality in $\triangle A'PB$, we obtain that

$$PA + PB = PA' + PB \geq A'B = P_*A' + P_*B = P_*A + P_*B,$$

showing that P_* is the required minimizer of $PA + PB$.

Solution to Exercise 2.5

No. For example, consider any triangle $\triangle ABC$, and draw a line parallel to the side BC such that it cuts AB and AC at *interior* points B', C'.

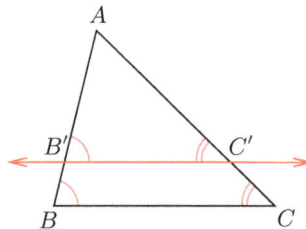

Then clearly $AB' < AB$, and so the triangles $\triangle ABC$ and $\triangle AB'C'$ are *not* congruent. However, all corresponding angles are equal:

$$\angle A = \angle A, \quad \angle ABC = \angle AB'C', \quad \angle ACB = \angle AC'B'.$$

So there is no "AAA Congruency Rule".

Solution to Exercise 2.6

By Exercise 1.20, each angle of a regular hexagon is $120°$. There are several ways of proceeding, and here is one. First construct the equilateral triangle $\triangle AOB$ with AB as base. With O as center, and radius AB, draw the circle $\mathcal{C}(O, AB)$. With B as center, and radius AB, draw a circle, which meets $\mathcal{C}(O, AB)$ at C. With C as center, and radius AB, draw a circle, which meets $\mathcal{C}(O, AB)$ at D. With D as center, and radius AB, draw a circle, which meets $\mathcal{C}(O, AB)$ at E. With E as center, and radius AB, draw a circle, which meets $\mathcal{C}(O, AB)$ at F. Then $ABCDEF$ is the required hexagon. Indeed, each of the triangles $\triangle ABO, \triangle BCO, \triangle CDO, \triangle DEO, \triangle EFO$ is equilateral, and so all the angles in each of these triangles is $60°$. It follows that $\angle FOA = 360° - 5 \cdot 60° = 60°$. Moreover, $\triangle FOA$ is isosceles, and so we conclude that $\angle OFA = \angle FAO = 60°$. By the SAS Congruency Rule, $\triangle FOA \simeq \triangle AOB$, and consequently, $AF = AB$, and $\triangle FOA$ is equilateral too. As all the sides in the hexagon $ABCDEF$ are equal, and each angle is $120°$, $ABCDEF$ is a regular hexagon.

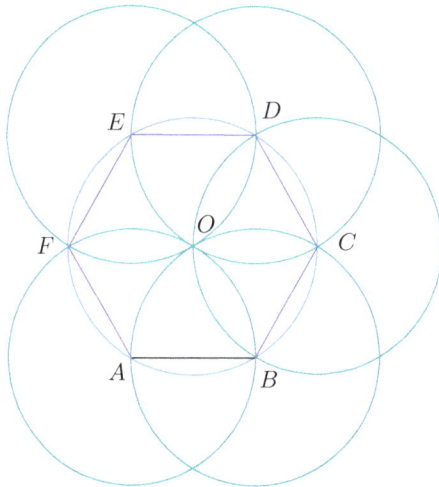

Solution to Exercise 2.7

We would be done if we manage to construct an angle of measure $1°$. We can first draw $19°$ ten times, so that when we compare with $180°$, we can get an angle of measure $10°$. This angle of $10°$ can be doubled to get an angle of $20°$. When we compare the angle of $20°$ with the given angle of $19°$, we obtain the required angle of $1°$. Done!

Another way of proceeding is to observe that $75° = 60° + 60°/4$ can be constructed, and with the data, also an angle of measure $4 \cdot 19° = 76°$ can be constructed. Again, we obtain $1° = 76° - 75°$.

Solution to Exercise 2.8

If all points in the plane are of the same colour, then we are done.

Otherwise there exists a pair of points of opposite colours, say A_1 (red) and B (blue). Let us assume that there does not exist a monochromatic equilateral triangle. (The plan is to arrive at a contradiction.) Look at the points A_2, A_3, A_4, A_5, A_6 completing the regular hexagon with center B and A_1 as the starting vertex.

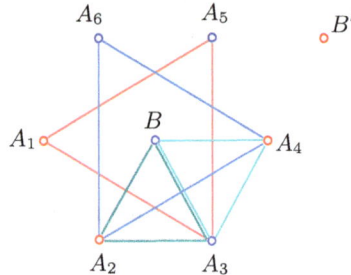

Among A_3 and A_5 at least one should be blue, since otherwise $\triangle A_1 A_3 A_5$ would be a monochromatic equilateral triangle. Without loss of generality, let us suppose that A_3 is blue. Reflect B in $A_4 A_5$ to obtain B'. Then A_4 must be red, since otherwise $\triangle B A_3 A_4$ would be a monochromatic equilateral triangle. Also, A_2 must be red, otherwise $\triangle A_2 A_3 B$ is a monochromatic equilateral triangle. But by looking at the equilateral triangle $\triangle A_2 A_4 A_6$, we conclude that A_6 must be blue. As $\triangle A_3 A_6 B'$ is equilateral, B' must be red. Then by looking at $\triangle A_5 A_4 B'$, we see that A_5 must be blue. But then $\triangle A_6 A_5 B$ is a monochromatic equilateral triangle, a contradiction. Hence our original assumption about the nonexistence of a monochromatic equilateral triangle must be false!

Solution to Exercise 2.9

Consider a line passing through P, intersecting the bounding rays of the given angle at the points A, A', such that P is the midpoint of AA'. Without loss of generality, assume that $\angle A' AO$ is acute. We will show that the area of the resulting triangle $\triangle OAA'$ is the smallest. To this end, consider a different line passing through P, meeting the bounding rays at B, B', and draw a line parallel to OA through the point A', meeting $B'P$ at the point B''. Then in $\triangle A'B''P$ and $\triangle ABP$, we have that $A'P = AP$, the opposite angles $\angle A'PB''$ and $\angle APB$ are equal, and the alternate angles $\angle PA'B''$ and $\angle PAB$ are equal. So by the ASA Congruency Rule, $\triangle A'B''P \simeq \triangle ABP$. From here, the claimed minimality follows since

$$
\begin{aligned}
\triangle B'OB &= \text{area}(OA'PB) + \triangle A'B''P + \triangle A'B'B'' \\
&= \text{area}(OA'PB) + \triangle ABP + \triangle A'B'B'' = \triangle A'OA + \triangle A'B'B'' \\
&\geq \triangle A'OA.
\end{aligned}
$$

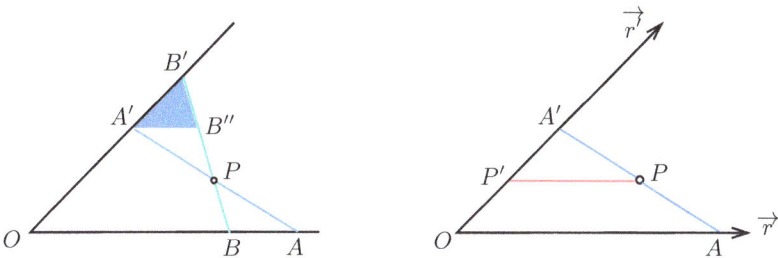

The construction: Draw a line through P parallel to one of the bounding rays $\overrightarrow{r}$, and suppose that it meets the other bounding ray $\overrightarrow{r'}$ at P'. Mark another point A' on $\overrightarrow{r'}$ such that $OP' = P'A'$. Join A' to P, and extend this to meet the ray $\overrightarrow{r}$ at A. AA' is the required line.

Reason: In triangle $\triangle A'OA$, P' is the midpoint of the side OA', and the line PP' is parallel to OA. So by the Midpoint Theorem 2, this line meets $A'A$ at its midpoint. Hence P bisects $A'A$.

Solution to Exercise 2.10

The fallacy is the purposely misleading diagram drawn, where the point of intersection O is shown to lie *inside* the triangle $\triangle ABC$, while in fact this point always lies *outside* the triangle, on the circumcircle of the triangle, except in the case when the triangle is isosceles or equilateral, in which case the angle bisector and the perpendicular bisector will coincide. (We will show this below, although we will make use of some results not yet shown. The reader can skip this now, and return to it again later.) If $AB < AC$, it can also be shown that the point P lies *outside* AB. Thus the proof outline still holds, except that we get $AB = AP - PB$ (instead of $AB = AP + PB$). So now there is no reason why AB must equal AC! The case when $AB > AC$ is analogous. Fallacy sorted.

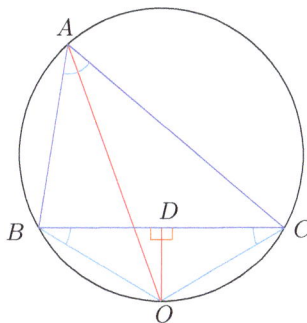

O lies on the circumcircle. Let the angle bisector of $\angle A$ meet the circumcircle at the point O. Drop a perpendicular from O to BC meeting BC in D. We'd like to show that $BD = DC$ (showing that OD is the perpendicular bisector of BC, meeting the angle bisector of $\angle A$ at O). An arc of a circle subtends the same angle in its opposite arc (Theorem 5.6). So $\angle OBC = \angle OAC$ $(=)$ $\angle BAO = \angle BCO$. Thus $\angle OBC = \angle BCO$, and so $\triangle OBC$ is isosceles, with $OB = OC$. By the AAS Congruency Rule, $\triangle OBD \simeq \triangle OCD$. Consequently $BD = DC$ (CPCT).

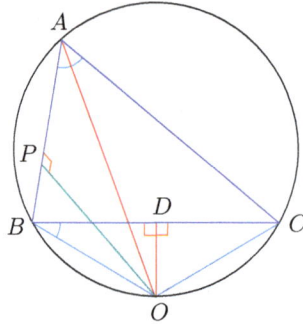

If $AB < AC$, then P lies outside AB. Suppose not. As $AB < AC$, $\angle B > \angle C$. Then

$$\angle APO = \angle ABO + \angle POB > \angle ABO = \angle ABC + \angle OBC = \angle B + \frac{\angle A}{2}$$
$$> \frac{\angle B + \angle C}{2} + \frac{\angle A}{2} = \frac{180°}{2} = 90°.$$

This shows that OP can't be perpendicular to AB, a contradiction.

Solution to Exercise 2.11

Let A, B be any two points in the plane. Construct the regular pentagon $ABCDE$, and let the extensions of the sides AE, CD meet at P.

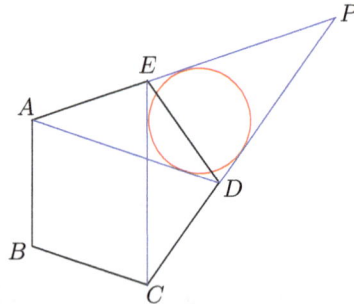

By symmetry, the points C, D are the reflections of the points A, E, respectively about the line BP. So if we consider the incircle of the triangle $\triangle ADP$, then upon reflection of this triangle about the angle bisector of $\angle P$ the incircle won't change. Hence $\triangle ADP$ and $\triangle CEP$ share their incircle. So

$$f(A) + f(D) = f(C) + f(E). \tag{6.6}$$

But thanks to the rotational symmetry of the pentagon about its center (by an angle of $72°$), it follows that

$$f(C) + f(E) = f(B) + f(D). \tag{6.7}$$

From (6.6) and (6.7), $f(A) + f(D) = f(C) + f(E) = f(B) + f(D)$, and cancelling $f(D)$ from the first and the last expressions, we have $f(A) = f(B)$. As the choice of A, B was arbitrary, we conclude that f is constant.

(Alternatively, start with the points A, A', and construct the regular hexagon $AA'BC'CB'$ as shown in the picture below. Suppose that $BC', B'C$ extended meet at D.

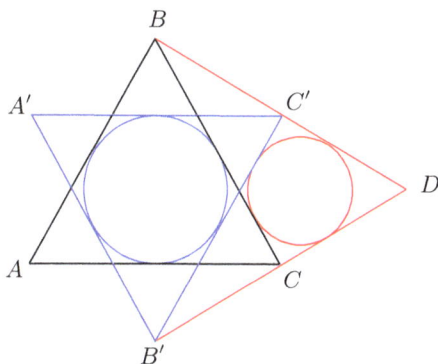

Then by the symmetry of the picture, we conclude that the points B', C' are the reflection of the points B, C with respect to the angle bisector of $\angle D$. If we consider the incircle of the triangle $\triangle BCD$, then upon reflection of this triangle about the angle bisector of $\angle D$, the incircle won't change. Hence $\triangle BCD$ and $\triangle B'C'D$ share their incircle. Thus $f(B) + f(C) + f(D) = f(B') + f(C') + f(D)$, and so

$$f(B) + f(C) = f(B') + f(C'). \tag{6.8}$$

We know that rotating a hexagon by $60°$ about its center produces no change, and so it follows that $\triangle ABC$ and $\triangle A'B'C'$ share their incircle. Hence

$$f(A) + f(B) + f(C) = f(A') + f(B') + f(C').$$

Using (6.8), we can cancel $f(B) + f(C)$ on the left-hand side with $f(B') + f(C')$ on the right-hand side, giving $f(A) = f(A')$.)

Solutions to the exercises from Chapter 3

Solution to Exercise 3.1

If a number n is composite, then we can find $q, d > 1$ such that $n = q \cdot d$, and so we can arrange q rows of d dots to form a rectangular array with totally $q \cdot d = n$ dots. For example, when $n = 12$, we have the following rectangular arrays:

$$
\begin{array}{c}
\quad\quad 4 \\
3 \; \begin{array}{cccc} \circ & \circ & \circ & \circ \\ \circ & \circ & \circ & \circ \\ \circ & \circ & \circ & \circ \end{array}
\end{array}
\qquad\qquad
\begin{array}{c}
\quad\quad 6 \\
\begin{array}{cccccc} \circ & \circ & \circ & \circ & \circ & \circ \\ \circ & \circ & \circ & \circ & \circ & \circ \end{array} \; 2
\end{array}
$$

But if a number p is prime, then we can't arrange the p numbers in a rectangular array, since otherwise, if there were q rows of d dots with q and d both > 1, then p would be divisible by $d \notin \{1, p\}$, a contradiction to the primeness of p!

$$
\begin{array}{c}
\quad\quad 7 \\
1 \; \begin{array}{ccccccc} \circ & \circ & \circ & \circ & \circ & \circ & \circ \end{array}
\end{array}
$$

Solution to Exercise 3.2

("If" part.) Let the diagonals AC, BD intersect at the point O. Since O bisects both AC and BD, in $\triangle AOD$ and $\triangle COB$, we have that $AO = OC$ and also $DO = OB$. Also the opposite angles $\angle AOD = \angle COB$. Thus by the SAS Congruency Rule $\triangle AOD \simeq \triangle COB$. Hence $\angle ADO = \angle CBO$ (CPCT), and as these are alternate angles for the transversal intersecting AD and BC, we conclude that $AD \parallel BC$. Also $AD = BC$ (CPCT). Since the opposite sides AD, BC are equal and parallel, it follows that $ABCD$ is a parallelogram.

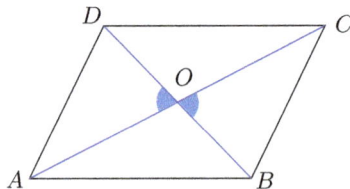

("Only if" part.) Let $ABCD$ be a parallelogram, with diagonals AC, BD intersecting in O. As $AD \parallel BC$, alternate interior angles to a transversal are equal, and so $\angle DAO = \angle BCO$ and $\angle ODA = \angle OBC$. Also $AD = BC$ (being opposite sides in a parallelogram). So $\triangle AOD \simeq \triangle COB$ by the ASA Congruency Rule. Hence $DO = BO$ and $AO = CO$ (CPCT), that is, O bisects AC and BD.

Solution to Exercise 3.3

Reflect the vertex A of triangle $\triangle ABC$ through the midpoint D of the side BC in order to obtain the point A' as shown. Then $ABA'C$ is a parallelogram (since the diagonals AA' and BC bisect each other at D).

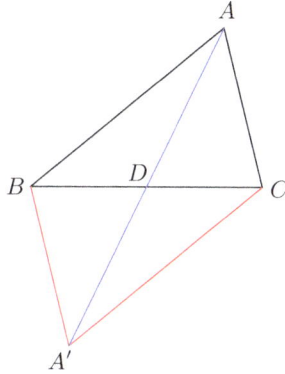

Using the triangle inequality, we have $AD = \dfrac{AA'}{2} < \dfrac{AB + BA'}{2} = \dfrac{AB + CA}{2}$.

Solution to Exercise 3.4

Let the centers of the three circles be C_1, C_2, C_3, their common intersection point be O, and the other intersection points be A_{12}, A_{23}, A_{31}.

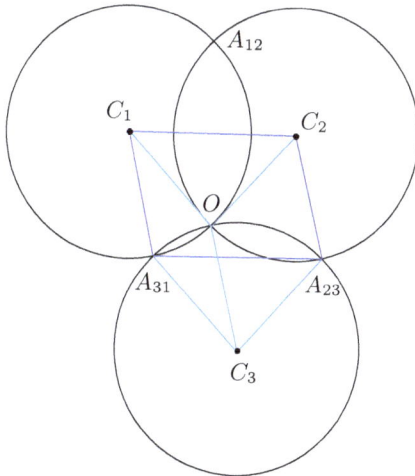

Then if each circle has radius R, we have $OC_1 = OC_2 = OC_3 = R$, and so

$\triangle C_1 C_2 C_3$ has circumradius R (and circumcenter O). So we will be done if we manage to show that $\triangle A_{23} A_{31} A_{12} \simeq \triangle C_1 C_2 C_3$. We will use the SSS Rule to show this. Also, we will just establish that $C_1 C_2 = A_{23} A_{31}$, since the other equalities of the lengths of the sides are analogous. Note that the quadrilateral $A_{31} C_3 O C_1$ is a rhombus since each side equals R. In particular, $A_{31} C_1 \parallel C_3 O$. Similarly, $A_{23} C_3 O C_2$ is also a rhombus, again since each of its sides equals R. From Proposition 3.1, we know that a rhombus is a paralellogram. Hence we have $A_{23} C_2 \parallel C_3 O$. But as $A_{31} C_1 \parallel C_3 O$ too, we conclude that $A_{31} C_1 \parallel A_{23} C_2$. Moreover, they both have length R. Thus in the quadrilateral $C_1 C_2 A_{23} A_{31}$, the opposite sides $A_{31} C_1$ and $A_{23} C_2$ are equal and parallel, showing that it is a parallelogram. So $C_1 C_2 = A_{23} A_{31}$.

Solution to Exercise 3.5

Let M be any point in the plane. Consider a square $ABCD$ whose center is M. Let the line through M parallel to AB meet AD, BC at D', B', respectively. Also, let the line through M parallel to AD meet AB, CD at A' and C', respectively.

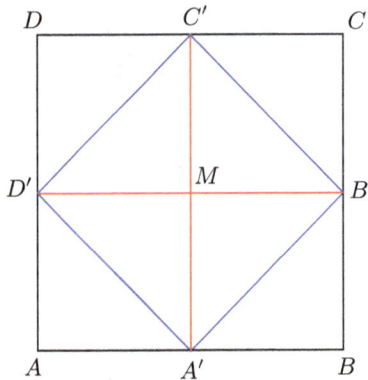

Then by considering the four small squares having M as a common vertex, we have

$$\begin{aligned}
-4f(M) &= (f(A') + f(A) + f(D')) + (f(B') + f(B) + f(A')) \\
&\quad + (f(C') + f(C) + f(B')) + (f(D') + f(D) + f(C')) \\
&= 2(f(A') + f(B') + f(C') + f(D')) \\
&\quad + (f(A) + f(B) + f(C) + f(D)) \\
&= 2 \cdot 0 + 0 = 0,
\end{aligned}$$

where we have also used the facts that $ABCD$ and $A'B'C'D'$ are squares. Hence $f(M) = 0$. But the choice of the point M was arbitrary, and so $f \equiv 0$.

Solution to Exercise 3.6

Call the given quadrilateral $ABCD$, and let P, Q, R, S be the midpoints of AB, BC, CD, DA, respectively. Join A to C. By the Midpoint Theorem in $\triangle ACD$, the line RS joining the midpoints is parallel to AC, and half its length. Similarly, again by the Midpoint Theorem in $\triangle ABC$, PQ is parallel to AC and half its length. Thus in quadrilateral $PQRS$, the pair of opposite sides RS and PQ are equal (both being half of AC) and are parallel (as they are both parallel to AC). Hence $PQRS$ is a parallelogram.

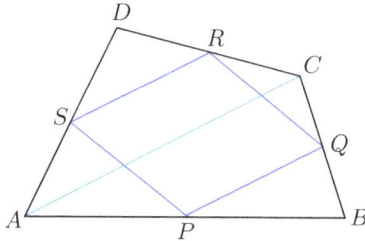

Solution to Exercise 3.7

Let F, D, E be the midpoints of the sides AB, BC, CA, respectively.

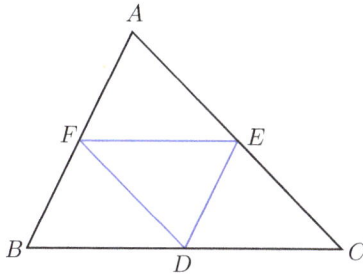

By the Midpoint Theorem,

$$FE = \frac{1}{2}BC = BD = DC,$$

$$DE = \frac{1}{2}AB = AF = BF,$$

$$DF = \frac{1}{2}AC = AE = CE.$$

Thus by the SSS Congruency Rule, $\triangle AFE \simeq \triangle FBD \simeq \triangle EDC \simeq \triangle DEF$.

Solution to Exercise 3.8

$DS = AD/2 = BC/2 = BQ$, and $DS \parallel BQ$. Thus $BQDS$ is a parallelogram. See the picture on the left below. Hence $BS \parallel DQ$. Let BS, DQ meet AC in M, N respectively. In $\triangle AND$, $SM \parallel ND$ and S is the midpoint of AD. By Midpoint Theorem 2, M is the midpoint of AN. Similarly, applying Midpoint Theorem 2 in $\triangle CMB$, we obtain that N is the midpoint of CM. Consequently $AM = MN = NC$, that is, AC is trisected by BS and DQ.

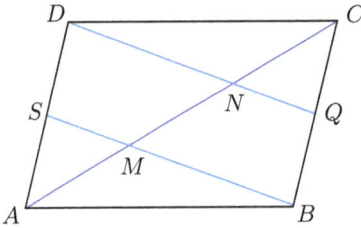

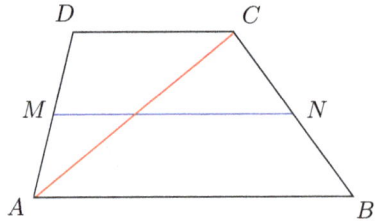

Solution to Exercise 3.9

See the picture on the right above. Suppose that MN intersects the diagonal AC in the point O, and let O' be the midpoint of AC. By the Midpoint Theorem in $\triangle ACD$, $MO' \parallel CD$. But as $AB \parallel CD$, $MO' \parallel AB$ as well. Also, by the Midpoint Theorem in $\triangle ABC$, $NO' \parallel AB$. Thus M, O', N are collinear. (Otherwise we would have *two* parallel lines to AB passing through O', contradicting the Parallel Postulate!) Thus $O = O'$. By the Midpoint Theorem in $\triangle ACD$ and $\triangle ABC$,

$$MO = \frac{CD}{2} \text{ and } ON = \frac{AB}{2},$$

and so $MN = MO + ON = \dfrac{CD}{2} + \dfrac{AB}{2} = \dfrac{AB + CD}{2}.$

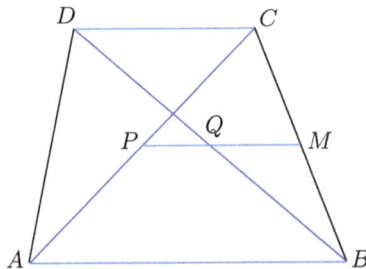

Solution to Exercise 3.10

Let the trapezium be $ABCD$, where we may assume without loss of generality[1] that $AB > CD$, and let P, Q be the midpoints of the diagonals AC, BD, respectively. Join P to the midpoint M of the side BC. By the Midpoint Theorem in $\triangle ABC$, PM is parallel to AB. Since $AB \parallel CD$, PM is also parallel to CD. In $\triangle BCD$, M is the midpoint of BC, and $PM \parallel CD$, and so by the Midpoint Theorem 2, PM meets BD in the midpoint of BD, namely Q. Thus PQ is parallel to CD and also to AB. By Midpoint Theorem 2 applied to $\triangle ABC$ and $\triangle BCD$, we obtain

$$PM = \frac{AB}{2} \text{ and } QM = \frac{CD}{2}.$$

Consequently,

$$PQ = PM - QM = \frac{AB - CD}{2}.$$

Solution to Exercise 3.11

To use the picture below, we make sure that the altitude AD is inside the triangle, and so we will label the triangle appropriately to arrange this. Choose A to be the vertex so that $\angle A$ is the largest angle amongst the angles of $\triangle ABC$. Drop a perpendicular from the vertex A to the side BC, meeting it at D. Then construct the midpoints F, E of the sides AB, AC, respectively, and finally join D to F, and D to E. The claim is that $\triangle BFD, \triangle AFD, \triangle AED$ and $\triangle DEC$ are all isosceles. Join F to E, and let EF meet AD in G.

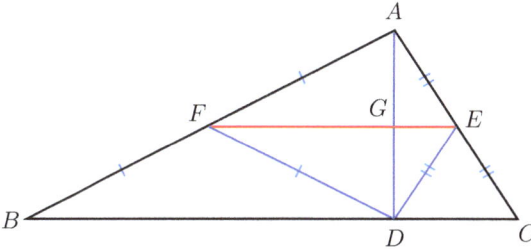

Then by the Midpoint Theorem, $EF \parallel BC$, and since $AD \perp BC$, it follows that $AD \perp EF$. By Midpoint Theorem 2 applied in $\triangle ABD$, we conclude that G is the midpoint of AD. By the SAS Congruency Rule, $\triangle AFG \simeq \triangle DFG$. Hence $AF = FD$ (CPCT). Hence $\triangle AFD$ is isosceles. Also, $BF = AF = FD$, and so $\triangle BFD$ is isosceles too. Similarly, we can show that $\triangle AED$ and $\triangle DEC$ are isosceles as well.

---

[1]If $AB = CD$, then $ABCD$ is a parallelogram, and the diagonals bisect each other, so that the midpoints of the two diagonals are coincident, and there is nothing to prove.

Solution to Exercise 3.12

Consider the triangles in two consecutive steps as shown. We will prove that their centroids coincide.

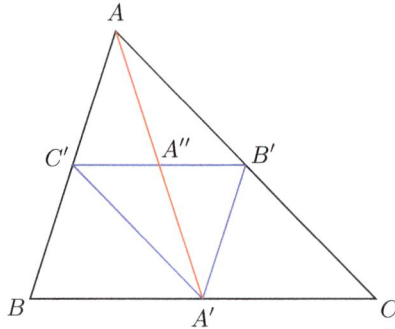

It is enough to show that if AA' intersects $B'C'$ at A'', then A'' bisects $B'C'$. By the Midpoint Theorem, $B'C'$ is parallel to BC. Also in $\triangle ABA'$, C' is the midpoint of AB and $C'A''$ is parallel to BA', and so by Midpoint Theorem 2, it follows that A'' is the midpoint of AA' and $C'A'' = BA'/2 = BC/4$. Similarly, by considering $\triangle ACA'$, we know that $A''B'$ is the line segment joining the midpoints, and so by the Midpoint Theorem, $A''B' = A'C/2 = BC/4$. Thus we have shown that $C'A'' = BC/4 = A''B'$, that is, A'' bisects $B'C'$. Hence $A'A''$ is a median of $\triangle A'B'C'$. Similarly BB', CC' are also extensions of medians of $\triangle A'B'C'$. So the triangles $\triangle ABC$ and $\triangle A'B'C'$ have the same centroid.

When one goes far down the sequence of triangles, the triangles become smaller and smaller, and there is very little distance between the vertices of the triangle $\triangle A_n B_n C_n$ and its centroid. Hence the centroids must converge to P too. As the centroid position is preserved at each step, it follows that P must be the centroid of the original triangle $\triangle ABC$.

This is also expected physically. If we imagine unit masses, say 1 kg, placed at A, B, C, then by writing each vertex mass as $0.5 + 0.5$ kg, we can instead place 1 kg at the midpoints of the sides, because each side with mass 0.5 kg at either end can be replaced by a mass of 1 kg at the midpoint of the respective side. So instead of 1 kg weights at the points A, B, C, we now have equivalently 1 kg weights at A', B', C'. The process can be continued, and eventually we get a 3 kg weight at the "limiting" point of the sequence of nested triangles. On the other hand, the mass distribution at the three vertices of $\triangle ABC$ could also be reduced to an equivalent form by first replacing the masses of 1 kg at B, C by a mass of 2 kg at the midpoint A' of BC, and hence arriving at the equivalent system of having masses of 1 kg and 2 kg at A, A', respectively. But then this can be replaced by a mass of 3 kg at the point G which divides AA' (a median in $\triangle ABC$) in the ratio $2:1$ ($= AG : A'G$), which is the centroid of $\triangle ABC$. So the limiting point of the sequence of triangles must be the centroid!

Solution to Exercise 3.13

Draw the angle bisector of $\angle A$, and let it meet BC in D'.

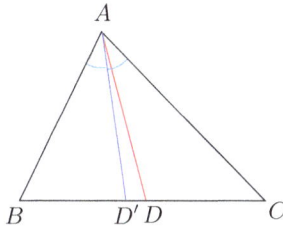

Then by the Angle Bisector Theorem, $AB : AC = BD' : D'C$. But we have been given that $AB : AC = BD : DC$. Hence $BD' : D'C = BD : DC$. Adding one to both sides gives

$$\frac{BC}{D'C} = \frac{BC}{DC},$$

that is, $D'C = DC$, and so D, D' are coincident. Hence $AD' = AD$ is the angle bisector of $\angle A$.

Solution to Exercise 3.14

Let $ABCD$ be the parallelogram, and AB be the chosen base. Draw the diagonal AC.

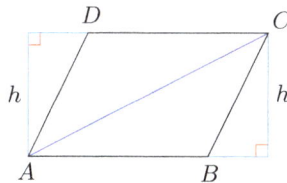

Then the parallelogram is divided into the two congruent triangles $\triangle ABC$ and $\triangle CDA$. If h is the distance between the parallel lines containing the segments AB, CD, then we have $\triangle ABC = (AB \cdot h)/2$. Consequently, the area of the parallelogram $ABCD$, being twice the area of $\triangle ABC$, is equal to $AB \cdot h$.

Solution to Exercise 3.15

As $AB : BC : CA = 4 : 5 : 6$, there exists a $\lambda > 0$ such that $AB = 4\lambda$, $BC = 5\lambda$ and $CA = 6\lambda$.

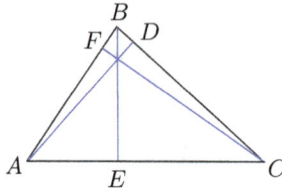

We have that $\triangle ABC = \dfrac{BC \cdot AD}{2} = \dfrac{AC \cdot BE}{2} = \dfrac{AB \cdot CF}{2}$. So

$$AD = \frac{2\triangle ABC}{\lambda} \cdot \frac{1}{5}, \quad BE = \frac{2\triangle ABC}{\lambda} \cdot \frac{1}{6}, \quad CF = \frac{2\triangle ABC}{\lambda} \cdot \frac{1}{4},$$

that is, $CF : AD : BE = \dfrac{1}{4} : \dfrac{1}{5} : \dfrac{1}{6}$.

Solution to Exercise 3.16

Let I be the incenter, and join I to the three vertices of $\triangle ABC$.

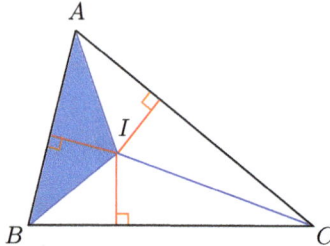

Then altitude dropped from I to the side BC has length r since the incenter (being a point on the angle bisectors of the three angles) is equidistant from the three sides. Thus the area of $\triangle ABI$ is $(AB \cdot r)/2$.
Hence

$$\Delta = \triangle ABC$$

$$= \triangle ABI + \triangle BCI + \triangle CAI$$

$$= \frac{c \cdot r + a \cdot r + b \cdot r}{2}$$

$$= r \cdot \frac{a + b + c}{2}.$$

Rearranging, we obtain $r = \dfrac{2\Delta}{a + b + c}$.

Solution to Exercise 3.17

If h_A, h_B, h_C are the altitudes dropped from the vertices A, B, C, respectively, to their opposite sides, then we have

$$\triangle ABC = \frac{AB \cdot h_C}{2} = \frac{BC \cdot h_A}{2} = \frac{CA \cdot h_B}{2}.$$

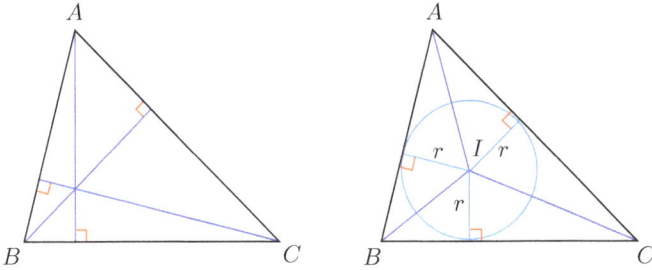

On the other hand, from Exercise 3.16, $\triangle ABC = \dfrac{AB + BC + CA}{2} \cdot r$. So

$$
\begin{aligned}
\frac{1}{r} &= \frac{AB + BC + CA}{2 \triangle ABC} \\
&= \frac{AB}{2 \triangle ABC} + \frac{BC}{2 \triangle ABC} + \frac{CA}{2 \triangle ABC} \\
&= \frac{1}{h_C} + \frac{1}{h_A} + \frac{1}{h_B}.
\end{aligned}
$$

Solution to Exercise 3.18

The fallacy is revealed by noticing that in fact there is a gap (of unit square area!) in the 5×13 rectangle between the pieces, sinces the "slopes" of the trapezium piece and the big triangular piece don't quite match, and $\angle ACB \neq \angle A'C'B'$. See the following picture.

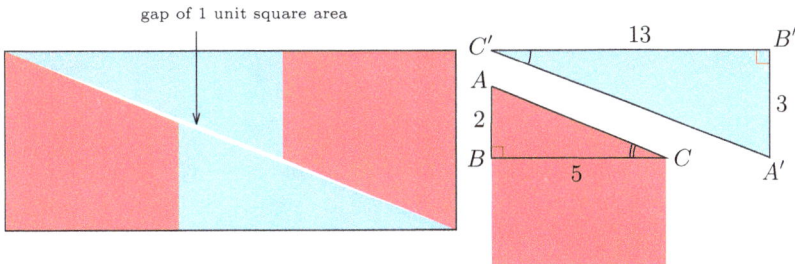

That $\angle ACB \neq \angle A'C'B'$ can be justified using triangle similarity as follows. (We will learn about similar triangles in Chapter 4.) As

$$\frac{AB}{A'B'} = \frac{2}{3} \neq \frac{5}{13} = \frac{BC}{B'C'},$$

we know that $\triangle ABC$ is not similar to $\triangle A'B'C'$. But angle $\angle B = 90° = \angle B'$, and so by the AA Similarity Rule, no other pair of corresponding angles can be equal! Hence $\angle ACB \neq \angle A'C'B'$.

Solution to Exercise 3.19

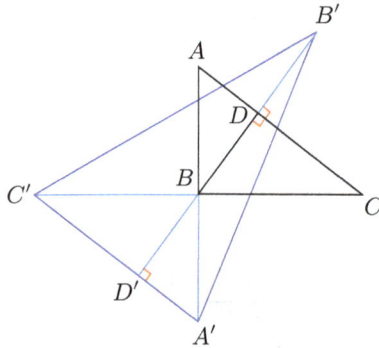

We note that B, D, B' are collinear and so $BB' = 2BD$. Also, $\triangle A'BC'$ is the rotation of $\triangle ABC$ by $180°$. Hence the altitude BD is also rotated through $180°$ to get the BD'. Hence D', B, B' are collinear and $D'B = BD$. So the altitude $B'D'$ to the side $A'C'$ in $\triangle A'B'C'$ has length $B'D' = BD + BB' = BD + 2BD = 3BD$. Hence

$$\triangle A'B'C' = \frac{1}{2}A'C' \cdot B'D'$$
$$= \frac{1}{2}AC \cdot (3BD) = 3 \cdot \frac{1}{2}AC \cdot BD$$
$$= 3\triangle ABC.$$

Solution to Exercise 3.20

The area of the staircase region is the sum of the areas of the small vertical rectangles, that is,

$$1 \cdot 1 + 1 \cdot 2 + 1 \cdot 3 + \cdots + 1 \cdot n = 1 + 2 + 3 + \cdots + n,$$

and the area of the big rectangle is $n \cdot (n+1)$.

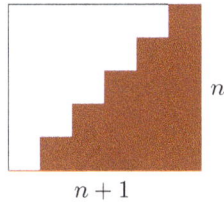

$n+1$

By the congruency of the shaded and the unshaded regions in the picture, we see that the area of the rectangle is twice the area of the staircase region. Thus we obtain $1 + 2 + \cdots + n = n(n+1)/2$.

Solution to Exercise 3.21

Let the equilateral triangle have a side length a, and let x, y, z be the distances of a point P inside the equilateral triangle to the three sides.

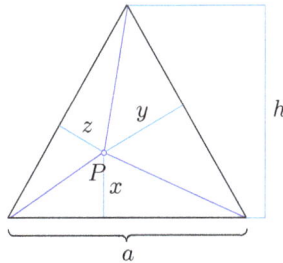

If h is the length of the altitude of the equilateral triangle, then the area of the equilateral triangle is $a \cdot h/2$. On the other hand, the area of the equilateral triangle is also the sum of the areas of the three smaller triangles obtained by joining P to the three vertices of the equilateral triangle. The altitudes in these smaller triangles, to the sides which are common with the equilateral triangle, have lengths x, y, z. Thus, $(a \cdot h)/2 = (a \cdot x)/2 + (a \cdot y)/2 + (a \cdot z)/2$, and so $x + y + z = h$.

An alternative "proof without words" is shown below.

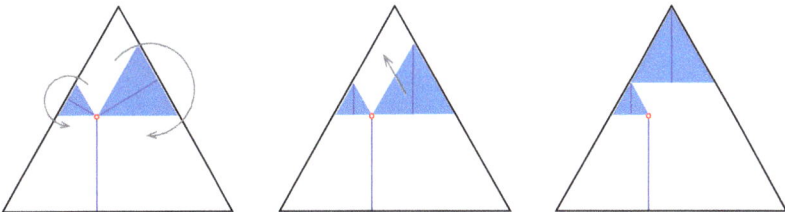

Solution to Exercise 3.22

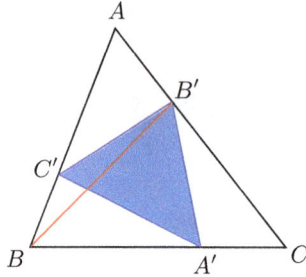

Case x = 1/2: Since $\triangle AB'C'$ and $\triangle AB'B$ share the altitude dropped from B', and because $AC' : AB = 2 : 3$, it follows that

$$\triangle AB'C' = \frac{2}{3}\triangle AB'B. \tag{6.9}$$

But $\triangle AB'B$ and $\triangle ABC$ share the altitude dropped from B. Also, we know that $AB' : AC = 1 : 3$. So

$$\triangle AB'B = \frac{1}{3}\triangle ABC. \tag{6.10}$$

From (6.9) and (6.10), $\triangle AB'C' = \frac{2}{3}\triangle AB'B = \frac{2}{3} \cdot \frac{1}{3}\triangle ABC = \frac{2}{9}\triangle ABC$.

General case: Similarly, $\triangle BC'A' = \frac{2}{9}\triangle ABC = \triangle CA'B'$. Hence

$$\triangle A'B'C' = \triangle ABC - (\triangle AB'C' + \triangle BC'A' + \triangle CA'B')$$
$$= \triangle ABC - 3 \cdot \frac{2}{9}\triangle ABC = \frac{1}{3}\triangle ABC.$$

Similarly, with $AC' : C'B = BA' : A'C = CB' : B'A = x$, then we get

$$\triangle AB'C' = \frac{x}{x+1}\triangle ABB' = \frac{x}{x+1} \cdot \frac{1}{x+1}\triangle ABC = \frac{x}{(x+1)^2}\triangle ABC,$$

and so $\triangle A'B'C' = \triangle ABC - 3 \cdot \frac{x}{(x+1)^2}\triangle ABC = \frac{x^2 - x + 1}{(x+1)^2}\triangle ABC.$

Particular case x = 0: When $x = 0$, $\triangle A'B'C'$ coincides with $\triangle ABC$, and the ratio of their areas must be 1. Our formula does deliver this:

$$\frac{x^2 - x + 1}{(x+1)^2}\bigg|_{x=0} = \frac{0^2 - 0 + 1}{(0+1)^2} = 1.$$

Particular case x = 1: When $x = 1$, A', B', C' are the midpoints of the sides of $\triangle ABC$, and then, using the Midpoint Theorem and the SSS Congruency Rule, we conclude that $\triangle AC'B' \simeq \triangle C'BA' \simeq \triangle B'A'C \simeq \triangle A'B'C'$. So the areas of these four little triangles are equal. But since these four areas add up to the area of $\triangle ABC$, it follows that the area $\triangle A'B'C'$ must be a fourth of the area of $\triangle ABC$. Our formula does indeed deliver this:

$$\frac{x^2 - x + 1}{(x+1)^2}\Big|_{x=1} = \frac{1^2 - 1 + 1}{(1+1)^2} = \frac{1}{4}.$$

Solution to Exercise 3.23

Let $a := BC$, $b := CA$, $c := AB$, and let s denote the semiperimeter. Then $AB + BD = s = DC + AC$, and so $BD = s - c$, $DC = s - b$.

Thus $\dfrac{BD}{DC} = \dfrac{s-c}{s-b}$. Similarly, $\dfrac{CE}{EA} = \dfrac{s-a}{s-c}$ and $\dfrac{AF}{FB} = \dfrac{s-b}{s-a}$.

Hence $\dfrac{BD}{DC} \cdot \dfrac{CE}{EA} \cdot \dfrac{AF}{FB} = \dfrac{s-c}{s-b} \cdot \dfrac{s-a}{s-c} \cdot \dfrac{s-b}{s-a} = 1$.

By the converse of Ceva's Theorem, AD, BE, CF are concurrent.

Solution to Exercise 3.24

Take a line segment OA with the given unit length. Draw a line segment AA' of unit length such that $\angle OAA' = 90°$. Then by Pythagoras's Theorem, we have $(OA')^2 = OA^2 + (AA')^2 = 1^2 + 1^2 = 2$, and so $OA' = \sqrt{2}$.

Now that we have got $\sqrt{2}$, we can construct $\sqrt{3}$ by considering the right angled triangle with nonhypotenuse side lengths $\sqrt{2}$ and 1. To this end, draw a line segment $A'A''$ of unit length such that $\angle OA'A'' = 90°$. Again by Pythagoras's Theorem, $(OA'')^2 = (OA')^2 + (A'A'')^2 = (\sqrt{2})^2 + 1^2 = 2 + 1 = 3$, and so $OA'' = \sqrt{3}$.

In general, having obtained $\sqrt{n}$, we can construct $\sqrt{n+1}$ by constructing a right angled triangle with the nonhypotenuse side lengths $\sqrt{n}$ and 1. Then the length of the hypotenuse will be $\sqrt{(\sqrt{n})^2 + 1^2} = \sqrt{n+1}$. Hence $\sqrt{n}$, $n \in \mathbb{N}$, can be obtained ("inductively") by constructing the "square root spiral" as shown below.

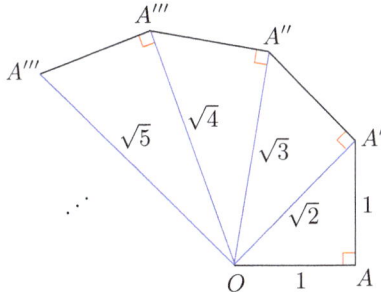

Solution to Exercise 3.25

We can first calculate the length ℓ of the diagonal of the rectangular base by using Pythagoras's Theorem: $\ell^2 = 4^2 + 12^2$.

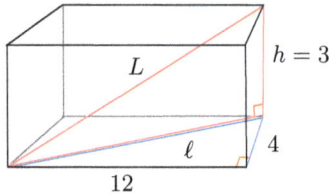

Next, we can look at the right angled triangle with the right angle included by sides of length ℓ and the height h of the box, and note that the required length L of the diagonal is the length of the hypotenuse of this right angled triangle, as shown in the picture. Thus by Pythagoras's Theorem again,

$$L^2 = \ell^2 + h^2 = 4^2 + 12^2 + 3^2 = 16 + 144 + 9 = 169 = 13^2.$$

Consequently, $L = 13$ inches.

Solution to Exercise 3.26

If we imagine the cup made of paper, then we can cut away the circular base. If we made a vertical cut along the vertical line which passes through the point of location of the ant, and then spread out the paper, we get a rectangle, whose height is 5 inches, and whose base width is the circumference of the circular base, $2\pi r = 2\pi \cdot (12/\pi) = 24$ inches.

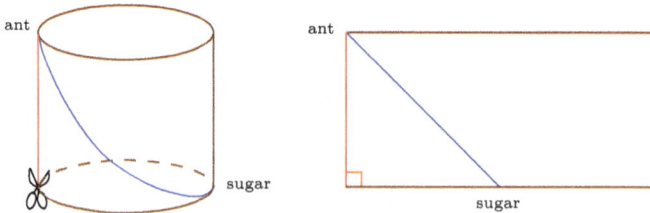

The ant is at one end of this rectangle, while the sugar grain is at the midpoint of the base, as shown above. Thus the length of the shortest path is given by the length of the line segment joining these two points, which, by Pythagoras's Theorem is

$$\sqrt{5^2 + \left(\frac{24}{2}\right)^2} = \sqrt{5^2 + 12^2} = \sqrt{25 + 144} = \sqrt{169} = 13 \text{ inches.}$$

Solution to Exercise 3.27

It is clear that since the sum of the two acute angles in a right angled triangle is $180° - 90° = 90°$, the triangles all "fit" at the corners. So the quadrilateral at the center is in fact one which has all angles equal to $90°$, and so it is "at worst" a rectangle. In fact it *is* a square, since all of its sides have length equal to $b - a$. Now the area of the big square is the sum of the areas of the four right triangles together with the area of the little square, giving the equality

$$c^2 = 4 \cdot \frac{ab}{2} + (b - a)^2,$$

and after an algebraic simplification, $c^2 = a^2 + b^2$. This proves Pythagoras's Theorem.

Solution to Exercise 3.28

(1) We have $5^2 = 25 = 9 + 16 = 3^2 + 4^2$. Also, $13^2 = 169 = 25 + 144 = 5^2 + 12^2$.

(2) If $c^2 = a^2 + b^2$, then $(nc)^2 = n^2 c^2 = n^2(a^2 + b^2) = n^2 a^2 + n^2 b^2 = (na)^2 + (nb)^2$. Furthermore, as a, b, c and n are natural numbers, so are na, nb and nc.

(3) If $c^2 = a^2 + b^2$, then it follows that the area of the square formed by the *overlapping* squares must be equal to the sum of the areas of the two small *unshaded* rectangles. The two rectangles are congruent, with side lengths $c - b$ and $c - a$, while the square formed by the overlapping squares has the side length $a + b - c$. Consequently, $(a + b - c)^2 = 2 \cdot (b - c)(c - a)$, and so

$$\frac{(b - c)(c - a)}{2} = \left(\frac{a + b - c}{2}\right)^2.$$

Note that $a + b - c > 0$, since otherwise $a + b \le c$ gives, upon squaring both sides, that $a^2 + b^2 + 2ab \le c^2 = a^2 + b^2$, that is, $2ab \le 0$, a contradiction. (This can also be seen geometrically, since if the two little squares don't overlap, they don't "cover" the area of the big square.) Moreover, $c^2 = a^2 + b^2 = (a + b)^2 - 2ab$, and so c is even if and only if $a + b$ is even. Hence $a + b - c$ is divisible by 2. Consequently, $(a + b - c)/2$ is a natural number.

(4) If a, b, c are integers, none of which is divisible by 3, then for some integers a', b', c', we can write $a = 3a' \pm 1$, $b = 3b' \pm 1$, $c = 3c' \pm 1$. If $a^2 + b^2 = c^2$, then $3 \cdot (\text{an integer}) + 2 = 1$, a contradiction.

(5) We have

$$(n^2 - m^2)^2 + (2nm)^2 = n^4 - 2n^2 m^2 + m^4 + 4n^2 m^2$$
$$= n^4 + 2n^2 m^2 + m^4$$
$$= (n^2 + m^2)^2.$$

As $n > m$, $n^2 - m^2 \in \mathbb{N}$. Thus $(2nm, n^2 - m^2, n^2 + m^2)$ is a Pythagorean triple.

Solution to Exercise 3.29

Let the hypotenuse be k, and the other two sides be the primes $p, p + 2$. Then we have $k^2 = p^2 + (p + 2)^2 = 2(p^2 + 2p + 2)$, showing that k^2, and hence also k must be even. Let $k = 2k'$. Then $p^2 = 2(k'^2 - p - 1)$, and so p must be even. As p is prime, $p = 2$. Thus $p + 2 = 2 + 2 = 4$, which is not prime; a contradiction.

Solution to Exercise 3.30

If we consider a layer of water at a certain height h from the bottom of the fish tank having a tiny thickness as compared to h, then the force on the walls having lengths a, b, c is $k_h \cdot a, k_h \cdot b, k_h \cdot c$ for some constant k_h depending on the height of the layer of water being considered. This force acts in a direction which is normal to the wall at the midpoint as shown below.

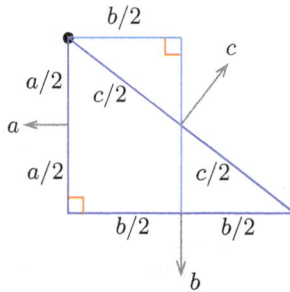

Recall that the rotational moment about an axis is the force multiplied by the distance of the line of the force to the axis, and so by the balancing of moments, and adding up all of these contributions at varying heights, we obtain the equation $K \cdot a \cdot (a/2) + K \cdot b \cdot (b/2) = K \cdot c \cdot (c/2)$, where K can be imagined to be the sum of all the k_h for varying heights h.

Solution to Exercise 3.31

In the picture below, $\triangle ABC$ is equilateral, and AD is the altitude dropped from A to the side BC.

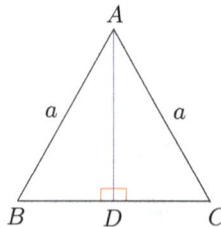

Then by the RHS Congruency Rule, $\triangle ADB \simeq \triangle ADC$ (since AD is common, $AB = AC$ and $\angle ADB = \angle ADC = 90°$). So $BD = BC$.

Hence $AD^2 = AB^2 - BD^2 = a^2 - \left(\dfrac{a}{2}\right)^2 = \dfrac{3}{4}a^2$, so that $AD = \dfrac{\sqrt{3}}{2}a$.

Moreover the area is $\triangle ABC = \dfrac{BC \cdot AD}{2} = \dfrac{a \cdot \frac{\sqrt{3}}{2}a}{2} = \dfrac{\sqrt{3}}{4}a^2$.

Solution to Exercise 3.32

The smallest distance among distinct points of a square lattice is the chosen unit length. However, the picture shows that if there is a regular hexagon with its vertices as lattice points, then there is a smaller regular hexagon whose vertices also lie on lattice points. Proceeding in this way, we can make the distance between distinct lattice points as small as we please, clearly a contradiction.

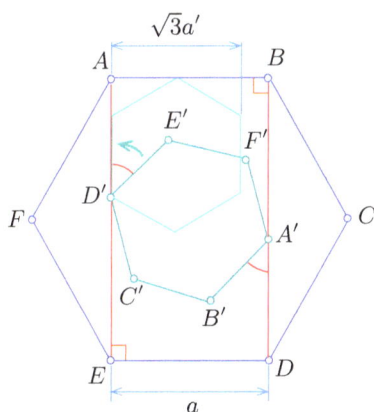

While the figure demonstrates quite clearly that the new hexagon has a strictly smaller side length, this can also be justified as follows. Note that the smaller hexagon has two of its vertices, A', D' lying on the sides AE, BD of the rectangle $AEDB$, and by the rotational symmetry present in the picture, it is also clear that $\angle AD'E' = \angle DA'B'$. So if we rotate the small hexagon about the point D' counterclockwise through the angle $\angle AD'E'$, we see that after rotation, $A'B'$ moves to become parallel to BD. Now if $a' := A'B'$ and $a := AB$, it follows that $\sqrt{3}a' = A'E' \leq AB = a$, that is $a' < a/\sqrt{3}$. Since $(1/\sqrt{3})^n$ goes to 0 as n increases without bound, we see that continuation of this process (of forming new smaller hexagons from older ones by a 90° rotation of the sides) will lead to smaller and smaller hexagons, with side lengths going to 0 — but as the vertices of these hexagons always lie on lattice points, this would mean that the distance between two distinct lattice points is not bounded below by 1, a contradiction.

Solution to Exercise 3.33

By Pythagoras's Theorem, $(ct_0)^2 + (vt)^2 = AB^2 + BB'^2 = AB'^2 = c^2t^2$, and so, by rearranging, $t = \dfrac{t_0}{\sqrt{1 - \dfrac{v^2}{c^2}}}$.

Solution to Exercise 3.34

Let x be the distance travelled by the snake (which is the same as that travelled by the peacock). Then we have the following picture.

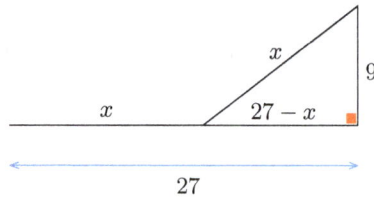

By Pythagoras's Theorem, $x^2 = (27 - x)^2 + 9^2$, and so $2 \cdot 27 \cdot x = 27^2 + 9^2$, that is, $x = 15$. Hence the distance to the snake's hole of the point they meet is $27 - x = 27 - 15 = 12$ cubits.

Solution to Exercise 3.35

From the picture below, $\Delta ADC \sim \Delta ACB \sim \Delta CDB$.

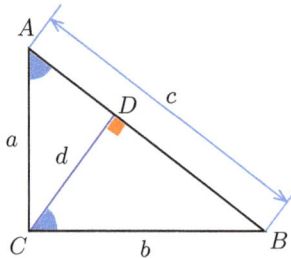

So with $c := AB$, we have $\dfrac{d}{a} = \dfrac{b}{c}$ and $\dfrac{d}{b} = \dfrac{a}{c}$. Consequently,

$$\frac{d^2}{a^2} + \frac{d^2}{b^2} = \frac{b^2}{c^2} + \frac{a^2}{c^2} = \frac{a^2 + b^2}{c^2} = \frac{c^2}{c^2} = 1.$$

Rearranging, we obtain $\dfrac{1}{a^2} + \dfrac{1}{b^2} = \dfrac{1}{d^2}$.

Solution to Exercise 3.36

Consider another triangle $\triangle A'B'C'$ with $A'B' = AB$, $A'C' = AC$ and $\angle A' = 90°$.

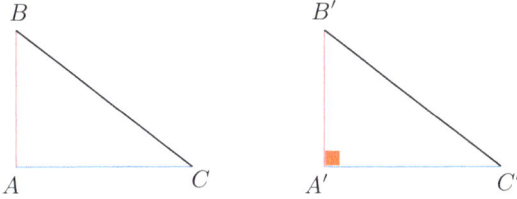

Then $\triangle A'B'C'$ is a right angled triangle, and so by Pythagoras's Theorem, $B'C'^2 = A'B'^2 + A'C'^2$. But $A'B'^2 + A'C'^2 = AB^2 + AC^2 = BC^2$. Thus we obtain $B'C'^2 = BC^2$, and so $B'C' = BC$. By the SSS Congruency Rule, $\triangle ABC \simeq \triangle A'B'C'$, and so $90° = \angle A' = \angle A$ (CPCT).

Solution to Exercise 3.37

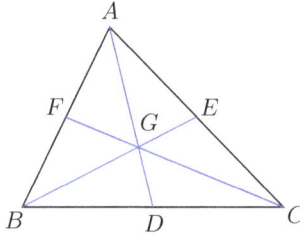

Since the centroid G divides the median AD in the ratio $2 : 1$, it follows from Apollonius's Theorem that

$$2CA^2 + 2AB^2 = 4AD^2 + BC^2 = 4\left(\frac{3}{2}GA\right)^2 + BC^2 = 9GA^2 + BC^2.$$

Similarly, by considering the medians through B and C, we also obtain

$$2AB^2 + 2BC^2 = 9GB^2 + CA^2,$$
$$2BC^2 + 2CA^2 = 9GC^2 + AB^2.$$

Adding the three equations above, we obtain, upon simplification, that

$$AB^2 + BC^2 + CA^2 = 3(GA^2 + GB^2 + GC^2).$$

Solution to Exercise 3.38

Let $x := OX$, $y := OY$, $z := OZ$, and $\xi := ZX$, $\eta := XY$, $\zeta := YZ$.

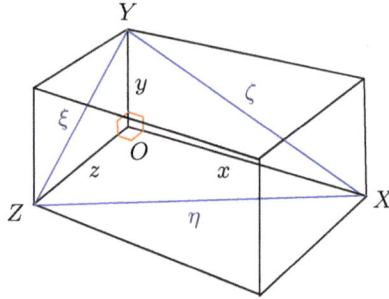

By Pythagoras's Theorem, $\zeta^2 = x^2 + y^2$, $\xi^2 = y^2 + z^2$, $\eta^2 = z^2 + x^2$. For the area of $\triangle XYZ$, we use Heron's Formula, which gives

$$
\begin{aligned}
16(\triangle XYZ)^2 &= (\xi + \eta + \zeta)(\xi + \eta - \zeta)(\xi - \eta + \zeta)(-\xi + \eta + \zeta) \\
&= \left((\xi + \eta)^2 - \zeta^2\right)\left(\zeta^2 - (\xi - \eta)^2\right) \\
&= -\zeta^4 + \left((\xi + \eta)^2 + (\xi - \eta)^2\right)\zeta^2 - (\xi + \eta)^2(\xi - \eta)^2 \\
&= -(\zeta^2)^2 + 2(\xi^2 + \eta^2)\zeta^2 - (\xi^2 - \eta^2)^2 \\
&= -(x^2 + y^2)^2 + 2(2z^2 + x^2 + y^2)(x^2 + y^2) - (y^2 - x^2)^2 \\
&= (x^2 + y^2)^2 + 4z^2(x^2 + y^2) - y^4 - x^4 + 2x^2y^2 \\
&= 4x^2y^2 + 4y^2z^2 + 4x^2z^2 \\
&= 16(\triangle OXY)^2 + 16(\triangle OYZ)^2 + 16(\triangle OZX)^2.
\end{aligned}
$$

Solutions to the exercises from Chapter 4

Solution to Exercise 4.1

Take any two equilateral triangles, and let them have side lengths a, a'. Then all the corresponding pairs of sides are in the same ratio $a : a'$, and all corresponding pairs of angles are equal (to $60°$). So the two equilateral triangles are similar.

Solution to Exercise 4.2

We check that the similarity relation is reflexive, symmetric and transitive:

(1) Reflexivity: A triangle $\triangle ABC$ is obviously similar to itself, since the corresponding sides are equal and the corresponding angles are equal:

$$AB = AB, \quad BC = BC, \quad CA = CA$$
$$\angle A = \angle A, \quad \angle B = \angle B, \quad \angle C = \angle C.$$

(2) Symmetry: If $\triangle ABC \sim \triangle A'B'C'$, then there exists a $k > 0$ such that

$$AB = k \cdot A'B', \quad BC = k \cdot B'C', \quad CA = k \cdot C'A'$$
$$\angle A = \angle A', \quad \angle B = \angle B', \quad \angle C = \angle C'.$$

Thus also $A'B' = k^{-1} \cdot AB$, $B'C' = k^{-1} \cdot BC$, $C'A' = k^{-1} \cdot CA$, and we know that $\angle A' = \angle A, \angle B' = \angle B, \angle C' = \angle C$. So $\triangle A'B'C' \sim \triangle ABC$.

(3) Transitivity: Let $\triangle ABC \sim \triangle A'B'C'$ and $\triangle A'B'C' \sim \triangle A''B''C''$. Then there exists a $k > 0$ such that

$$AB = k \cdot A'B', \quad BC = k \cdot B'C', \quad CA = k \cdot C'A'$$
$$\angle A = \angle A', \quad \angle B = \angle B', \quad \angle C = \angle C'$$

and there exists a $k' > 0$ such that

$$A'B' = k' \cdot A''B'', \quad B'C' = k' \cdot B''C'', \quad C'A' = k' \cdot C''A''$$
$$\angle A' = \angle A'', \quad \angle B' = \angle B'', \quad \angle C' = \angle C''.$$

Putting these together, we obtain also that

$$AB = (kk') \cdot A''B'', \quad BC = (kk') \cdot B''C'', \quad CA = (kk') \cdot C''A''$$
$$\angle A = \angle A'', \quad \angle B = \angle B'', \quad \angle C = \angle C'',$$

that is, $\triangle ABC \sim \triangle A''B''C''$.

Solution to Exercise 4.3

("If" part): Suppose that

$$\frac{AP}{PC} = \frac{BP}{PD}. \tag{6.11}$$

Draw a line through P which is parallel to AB, and suppose that it meets the side BC in the point Q.

Then by the Basic Proportionality Theorem, $\dfrac{AP}{PC} = \dfrac{BQ}{QC}$.

In light of (6.11), we now obtain $\dfrac{BP}{PD} = \dfrac{BQ}{QC}$ in $\triangle BCD$.

By the Basic Proportionality Theorem 2, $PQ \parallel CD$. Moreover, as $AB \parallel PQ$, and $PQ \parallel CD$, it follows that $AB \parallel CD$. Hence $ABCD$ is a trapezium with parallel sides AB, CD.

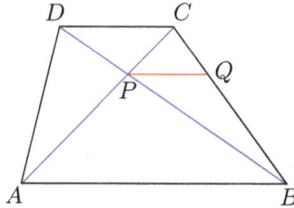

("Only if" part): Suppose that $ABCD$ is a trapezium with $AB \parallel CD$. Draw a line through P which is parallel to AB, meeting BC in Q. In $\triangle ABC$, $PQ \parallel AB$, and so by the Basic Proportionality Theorem,

$$\frac{AP}{PC} = \frac{BQ}{QC}. \tag{6.12}$$

In $\triangle BCD$, PQ is parallel CD (since PQ is parallel to AB, which is in turn parallel to CD), and so again by the Basic Proportionality Theorem,

$$\frac{BP}{PD} = \frac{BQ}{QC}. \tag{6.13}$$

By (6.12) and (6.13), $\dfrac{AP}{PC} = \dfrac{BQ}{QC} = \dfrac{BP}{PD}$.

Solution to Exercise 4.4

Drop perpendiculars PF and PE from P to the given lines. Extend PE, and let it meet the other line at B. Similarly, extend PF to meet the other line at C. Join BC. From P drop a perpendicular to BC to meet it at D. Join PD, and this is the required line. See the following picture.

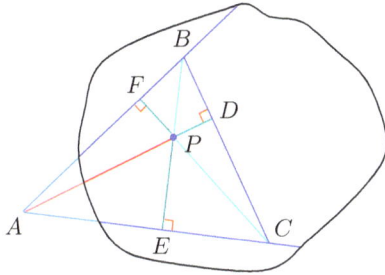

Rationale: CF and BE are two altitudes in $\triangle ABC$, and their intersection point P is the orthocenter. As the altitudes in a triangle are concurrent, it follows that the perpendicular dropped from P to BC must be the altitude from A to BC. So the line through P and D must pass through A!

An alternative method is the following. Drop perpendiculars from P to the two lines to meet them at O, O'. Through an arbitrarily chosen point $N \neq O$ on the line containing O, draw a line parallel to OO' to intersect the other line at N'. Erect perpendiculars at N, N', and let them meet at Q. PQ is the required line. See the picture on the left below.

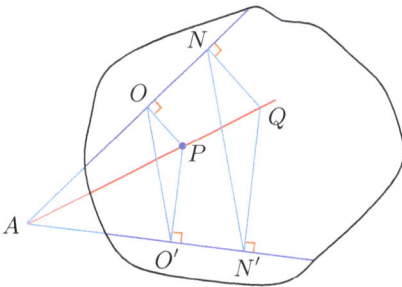

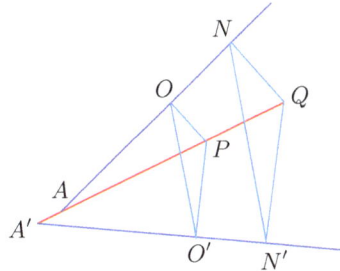

Rationale: Let PQ meet the two lines in A, A'. See the picture on the right above.

Then $\triangle ANQ \sim \triangle AOP$, and so $\dfrac{AQ}{PQ} = \dfrac{NQ}{NQ - OP}$.

Similarly, $\triangle A'N'Q \sim \triangle A'O'P$, and so $\dfrac{A'Q}{PQ} = \dfrac{N'Q}{N'Q - O'P}$.

Also, since $NN' \parallel OO'$ and because both NQ and OP are perpendicular to NO, it follows that $\angle N'NQ = \angle O'OP$. Similarly, $\angle NN'Q = \angle OO'P$. So $\triangle NN'Q \sim \triangle OO'P$.

So $\dfrac{NQ}{NQ - OP} = \dfrac{N'Q}{N'Q - O'P}$. Thus $\dfrac{AQ}{PQ} = \dfrac{A'Q}{PQ}$, and so A' and A are coincident.

Solution to Exercise 4.5

Let the line segment be AB, and let $\ell \parallel AB$. Take any point C as shown, so that the line segment AB and the point C lie on opposite sides of ℓ.

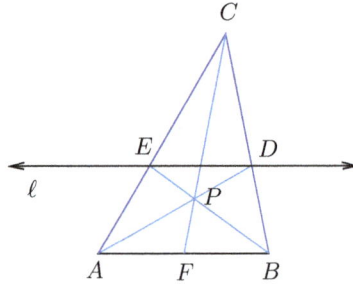

Join A to C, and B to C, and let them intersect ℓ in E, D, respectively. Also, let BE, AD meet in P. Join C to P, and extend it to meet AB in F. Then F is the midpoint of AB.

Justification: As $\ell \parallel AB$, $\dfrac{CE}{EA} = \dfrac{CD}{DB}$. By Ceva's Theorem, $\dfrac{\cancel{CE}}{\cancel{AE}} \cdot \dfrac{AF}{FB} \cdot \dfrac{\cancel{DB}}{\cancel{DC}} = 1$.

So $AF = FB$.

Solution to Exercise 4.6

See the picture below. If the side of the big square is 1, then the areas of the smaller rectangles and the smaller squares are obtained successively by halving the area of the previous rectangle or square. Hence these areas are precisely the terms $\frac{1}{2}$, $\left(\frac{1}{2}\right)^2$, $\left(\frac{1}{2}\right)^3, \cdots$ in geometric progression. As these areas add up to the area of the big square, we conclude that

$$\sum_{n=1}^{\infty} \left(\frac{1}{2}\right)^n = 1.$$

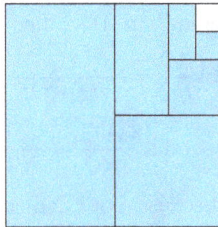

For the second identity, first note that if the big square has side length 1, then the shaded areas of the little squares in the leftmost picture are successively quartered forming the geometric progression $\frac{1}{4}$, $\left(\frac{1}{4}\right)^2$, $\left(\frac{1}{4}\right)^3, \cdots$, that is, $\left(\frac{1}{2}\right)^2$, $\left(\frac{1}{2}\right)^4$, $\left(\frac{1}{2}\right)^6, \cdots$.

Now to see that the shaded area is one-third of the area of the big square, we slide each shaded square to the right, as shown in the middle picture below. We could have also slid each square area upwards as shown in the rightmost picture.

Hence $3 \cdot$ (shaded area) $= 1$, that is, $\displaystyle\sum_{n=1}^{\infty} \left(\frac{1}{2}\right)^{2n} = \frac{1}{3}$.

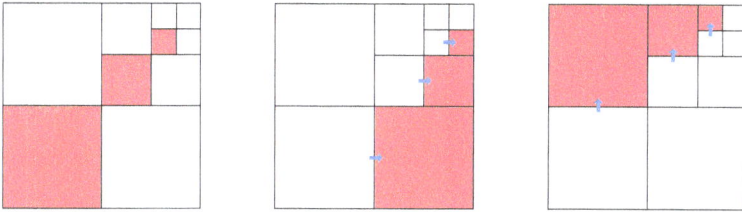

Solution to Exercise 4.7

Draw a ray $\overrightarrow{OA}$ and take $OA = 1$. Draw a ray $\overrightarrow{OB_1}$ which intersects the line containing OA only at O. Cut equal lengths

$$OB_1 = B_1B_2 = B_2B_3 = \cdots = B_{10}B_{11}$$

along the ray $\overrightarrow{OB_1}$. Join A to B_6, and construct $B_{11}P'$ parallel to B_6A, meeting OA (extended) at P'. Then $P' \equiv 11/6$. With O as center and radius OP', draw a circle which meets the line passing through O and A at the point $P \not\equiv P'$. Then $P \equiv -11/6$.

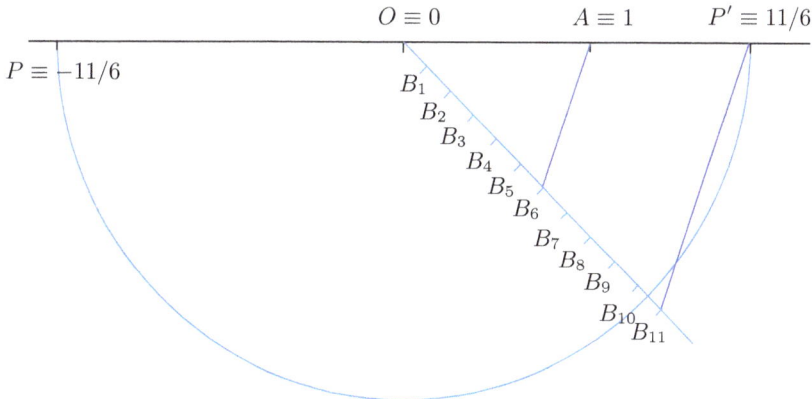

Solution to Exercise 4.8

Draw the external angle bisector at A, and let it meet the extension of BC in D'.

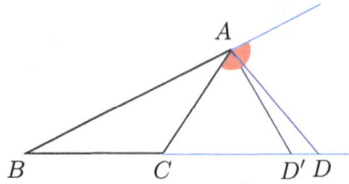

Then by the External Angle Bisector Theorem, $AB : AC = BD' : CD'$. But it is given that $AB : AC = BD : CD$. Hence $BD' : CD' = BD : CD$. Subtracting one from both sides gives $BC : CD' = BC : CD$, that is, $CD' = CD$, and so D, D' are coincident. Hence $AD' = AD$ is the external angle bisector at A.

Solution to Exercise 4.9

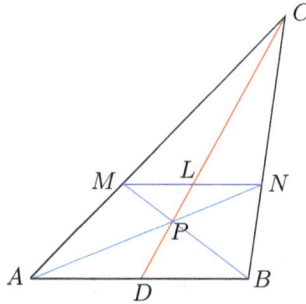

(1) Consider $\triangle CML$ and $\triangle CAD$. As $MN \parallel AB$ we have $\triangle CML \sim \triangle CAD$ by the AA Similarity Rule. Similarly $\triangle CMN \sim \triangle CAB$. Thus

$$\frac{ML}{AD} = \frac{CM}{AC} = \frac{MN}{AB}. \tag{6.14}$$

Consider the two triangles $\triangle MNP$ and $\triangle BAP$. As $MN \parallel AB$, we have $\triangle MNP \sim \triangle BAP$ by the AA Similarity Rule. Similarly $\triangle MLP \sim \triangle BDP$. Hence

$$\frac{ML}{BD} = \frac{MP}{PB} = \frac{MN}{AB}. \tag{6.15}$$

From (6.14) and (6.15) it follows that $\frac{MN}{AB} = \frac{ML}{AD} = \frac{ML}{BD}$, and so $AD = BD$.

Thus D is the midpoint of AB, and $AD = AB/2$. Substituting this fact in

$$\frac{MN}{AB} = \frac{ML}{AD},$$

we obtain $ML = MN/2$, that is, L bisects MN. This completes the proof.

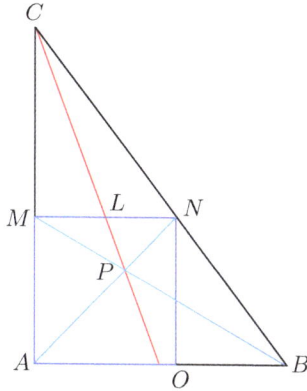

(2) Let the square be $AONM$ as shown. Extend AO to an arbitrary point B lying on the line passing through AO. Join B to N, and extend it to meet AM extended at the point C. Join BM and AN, and let them intersect in P. Join CP, and let it meet segment MN at L. Then L is the midpoint of MN. Reason: As $MN \parallel AB$, it follows from the previous part that L bisects MN.

Solution to Exercise 4.10

Let AC, BD meet at P, and let $u := AP$, $U := PC$, $v := BP$, $V := PD$.

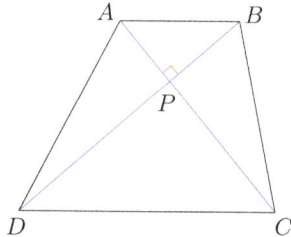

As $AB \parallel CD$, we have that $\angle PAB = \angle PCD$ and $\angle ABP = \angle PDC$.

By the AA Similarity Rule, $\triangle APB \sim \triangle CPD$, and so $\dfrac{u}{U} = \dfrac{AP}{PC} = \dfrac{BP}{PD} = \dfrac{v}{V}$.

By Pythagoras's Theorem in $\triangle DPA$, $\triangle APB$, $\triangle BPC$ and $\triangle CPD$,

$$AD^2 = u^2 + V^2, \tag{6.16}$$
$$AB^2 = u^2 + v^2, \tag{6.17}$$
$$BC^2 = v^2 + U^2, \tag{6.18}$$
$$CD^2 = U^2 + V^2. \tag{6.19}$$

Thus

$$(AD \cdot BC)^2 - (AB \cdot CD)^2 = (u^2 + V^2)(v^2 + U^2) - (u^2 + v^2)(U^2 + V^2)$$

$$= u^2 v^2 + \cancel{u^2 U^2} + \cancel{V^2 v^2} + U^2 V^2 - \cancel{u^2 U^2} - u^2 V^2 - v^2 U^2 - \cancel{v^2 V^2}$$

$$= u^2(v^2 - V^2) - U^2(v^2 - V^2) = U^2 V^2 \left(\frac{u^2}{U^2} - 1\right)\left(\frac{v^2}{V^2} - 1\right)$$

$$= U^2 V^2 \left(\frac{u^2}{U^2} - 1\right)^2 \tag{6.20}$$

$$\geq 0,$$

proving the first inequality. The second inequality is equivalent to the first, since

$$(AD + BC)^2 - (AB + CD)^2$$
$$= AD^2 + BC^2 - AB^2 - CD^2 + 2(AD \cdot BC - AB \cdot CD)$$
$$= u^2 + V^2 + v^2 + U^2 - (u^2 + v^2) - (U^2 + V^2) + 2(AD \cdot BC - AB \cdot CD)$$
$$= 0 + 2(AD \cdot BC - AB \cdot CD) = 2(AD \cdot BC - AB \cdot CD)$$

and the last expression is ≥ 0 if and only if $AD \cdot BC \geq AB \cdot CD$.

We claim that an equality in any of the inequalities holds if and only if $ABCD$ is a rhombus. The "if" part is obvious. Now we will show the "only if" part. Suppose that $AD \cdot BC = AB \cdot CD$. Then it follows from (6.20) that

$$U^2 V^2 \left(\frac{u^2}{U^2} - 1\right)^2 = 0,$$

and so $u = U$. Using the fact that $u/U = v/V$, it follows that $v = V$ as well. But then from (6.16)-(6.19), we obtain $AD = AB = BC = CD$.

Solution to Exercise 4.11

By the AA Similarity Rule, $\triangle ABC \sim \triangle A'B'C$, and so $\dfrac{s}{h} = \dfrac{d}{R_\oplus}$.

Consequently, $R_\oplus \approx 8(500) = 4000$ miles, that is $1.6 \times 4000 = 6400$ km.

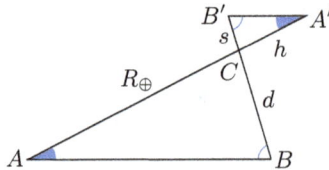

(In Astronomy, it is estimated that the radius of the Earth varies between 6353 and 6384 km, and the mean radius $R_\oplus$ is taken as 6371 km.)

Solution to Exercise 4.12

Let the line containing DE meet AB in F'.

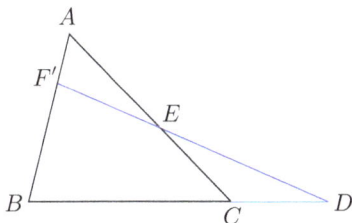

By Menelaus's Theorem, $\dfrac{BD}{DC} \cdot \dfrac{CE}{EA} \cdot \dfrac{AF'}{F'B} = 1$. Also $\dfrac{BD}{DC} \cdot \dfrac{CE}{EA} \cdot \dfrac{AF}{FB} = 1$ (given).

Hence $\dfrac{AF'}{F'B} = \dfrac{AF}{FB}$, and adding 1 to both sides, also $\dfrac{AB}{F'B} = \dfrac{AB}{FB}$.

Consequently $F'B = FB$ and so F, F' coincide.

Solution to Exercise 4.13

Join B to D, and suppose that the line ℓ cuts BD at T.

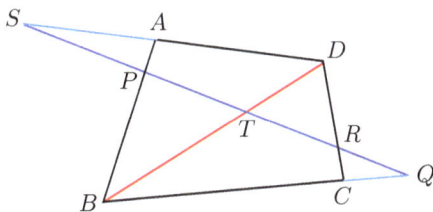

Then applying Menelaus's Theorem in $\triangle ABD$, we obtain $\dfrac{AP}{PB} \cdot \dfrac{BT}{TD} \cdot \dfrac{DS}{SA} = 1$.

Similarly, applying Menelaus's Theorem in $\triangle BCD$, we have $\dfrac{BQ}{QC} \cdot \dfrac{CR}{RD} \cdot \dfrac{TD}{BT} = 1$.

Mutltiplying the above, we obtain

$$1 = 1 \cdot 1 = \left(\frac{AP}{PB} \cdot \frac{BT}{TD} \cdot \frac{DS}{SA}\right) \cdot \left(\frac{BQ}{QC} \cdot \frac{CR}{RD} \cdot \frac{TD}{BT}\right) = \frac{AP}{PB} \cdot \frac{BQ}{QC} \cdot \frac{CR}{RD} \cdot \frac{DS}{SA}.$$

Solution to Exercise 4.14

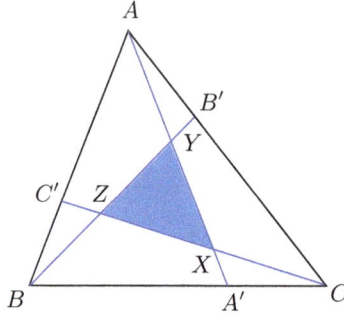

Menelaus's Theorem applied to $\triangle ABA'$ with CC' cutting it gives

$$1 = \frac{AC'}{C'B} \cdot \frac{BC}{A'C} \cdot \frac{A'X}{XA} = \frac{2}{1} \cdot \frac{3}{1} \cdot \frac{A'X}{XA},$$

giving $A'X : XA = 1 : 6$. As $AX + XA' = AA'$, it follows that $AA' = 7XA'$. Since $\triangle CXA'$ and $\triangle CAA'$ share the altitude dropped from C, and also $\triangle AA'C$ and $\triangle ABC$ share the altitude dropped from A, we obtain

$$\triangle XA'C = \frac{1}{7}\triangle AA'C = \frac{1}{7} \cdot \frac{1}{3}\triangle ABC.$$

Similarly, $\triangle AYB' = \triangle BZC' = \frac{1}{21}\triangle ABC$ too. Also,

$$\frac{1}{3}\triangle ABC = \triangle AA'C = \triangle AYB' + \text{Area}(B'YXC) + \triangle CXA'$$
$$= \frac{2}{21}\triangle ABC + \text{Area}(B'YXC),$$

and so $\text{Area}(B'YXC) = \frac{5}{21}\triangle ABC$. Similarly,

$$\text{Area}(AYZC') = \text{Area}(BZXA') = \frac{5}{21}\triangle ABC$$

as well. Putting all this together, we obtain

$$\triangle XYZ = \triangle ABC - 3 \cdot \frac{1}{21}\triangle ABC - 3 \cdot \frac{5}{21}\triangle ABC = \frac{1}{7}\triangle ABC.$$

Proceeding similarly, we can also treat the more general case when we have that $AC' : C'B = BA' : A'C = CB' : BA' = x$. We get $A'X : XA = 1 : (x(x+1))$, and so

$$\frac{\triangle XA'C}{\triangle ABC} = \frac{1}{(x^2+x+1)(x+1)}.$$

Moreover, the area of each of the small quadrilaterals, for example

$$\text{Area}(B'YXC) = \frac{x^2 + x - 1}{(x+1)(x^2 + x + 1)} \cdot \triangle ABC.$$

Finally, we obtain

$$\frac{\triangle XYZ}{\triangle ABC} = \frac{(x-1)^2}{x^2 + x + 1}.$$

When $x = 0$, we have $A' = B = Y$, $B' = C = Z$, $C' = A = X$, so that $\triangle XYZ = \triangle ABC$, and our formula delivers this:

$$\left.\frac{(x-1)^2}{x^2 + x + 1}\right|_{x=0} = \frac{(0-1)^2}{0^2 + 0 + 1} = 1.$$

When $x = 1$, A', B', C' are the midpoints of the sides of $\triangle ABC$. Consequently, the line segments AA', BB', CC' are the medians of $\triangle ABC$. But we know that the medians of a triangle are concurrent. Hence the points X, Y, Z all coincide (with the centroid of $\triangle ABC$). Thus the triangle $\triangle XYZ$ has collapsed to a point, and the ratio of the areas of $\triangle XYZ$ to that of $\triangle ABC$ is 0, and this agrees with our formula:

$$\left.\frac{(x-1)^2}{x^2 + x + 1}\right|_{x=1} = \frac{(1-1)^2}{1^2 + 1 + 1} = 0.$$

Solution to Exercise 4.15

(1) We first note that $\triangle APM$ and $\triangle AA'P$ are both isosceles, with the same base angle $\angle PAM$. So all the corresponding angles are equal, and so the two triangles are similar.

Hence $\dfrac{AM}{AP} = \dfrac{AP}{AA'}$, and as $AP = AB = AA'/2$, we obtain $AM = AB/2$.

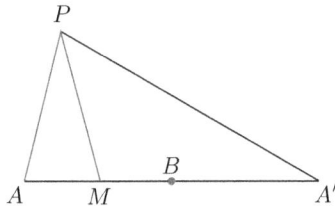

(2) With B as center and radius $AB =: r$, draw the circle $C(B,r)$. With A as center and radius r, draw the circle $C(A,r)$, and let it meet $C(B,r)$ at some point X_1. With X_1 as center and radius r, draw a circle which meets $C(B,r)$ at the point $X_2 \neq A$. With X_2 as center and radius r, draw a circle $C(X_2, r)$ which meets $C(B,r)$ at $A' \neq X_1$. Then we see that $AA' = 2r = 2AB$. (One way to see the first equality is to notice that each of the triangles $\triangle AX_1B$, $\triangle X_1BX_2$, $\triangle X_2BA'$ is equilateral, and so $\angle ABA' = 3 \cdot 60° = 180°$. Hence AA' is the diameter of $C(B,r)$.)

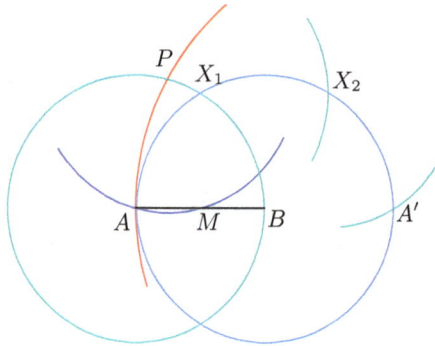

With A' as center and radius $AA' = 2r$, draw a circle $C(A', 2r)$ which meets $C(A, r)$ at a point P. With P as center and radius r, draw a circle $C(P, r)$ which meets AB in M. Then $PA = PM = r = AB$. Moreover, we also know that $PA' = AA' = 2r = 2AB$. By the previous part, $AM = AB/2$.

(3) In the above, we gave a procedure for constructing a point A' on the line containing AB such that $AA' = 2AB$. By now starting with the points B, A', we can construct a point A'' on the right of A' on the line containing A', B such that $BA'' = 2BA'$, so that $AA'' = 3AB$. And we can continue in this manner to obtain the point $A^{(n)}$ on the line containing A, B such that $AA^{(n)} = nAB$. Then with A as center and radius r, draw a circle $C(A, r)$. With $A^{(n)}$ as center and radius $AA^{(n)} = nr$, draw a circle $C(A^{(n)}, nr)$ which meets $C(A, r)$ at a point P. With P as center and radius r, draw a circle $C(P, r)$ which meets AB in M_n. Then proceeding in a manner akin to part (1), $AM_n = AB/n$.

Solution to Exercise 4.16

In $\triangle ABO$ and $\triangle OB'A'$, $\angle B = 90° = \angle B'$, and $\dfrac{AB}{OB'} = \dfrac{bc}{bd} = \dfrac{c}{d} = \dfrac{ac}{ad} = \dfrac{OB}{A'B'}$.

By the SAS Similarity Rule, $\triangle ABO \sim \triangle OB'A'$. Hence $\angle AOB = \angle OA'B'$. Thus

$$\angle AOA' = 180° - (\angle AOB + \angle A'OB')$$
$$= 180° - (\angle OA'B' + \angle A'OB') = 180° - 90° = 90°.$$

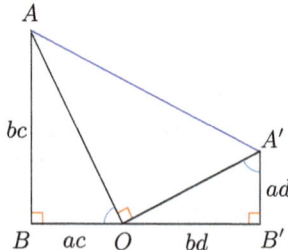

By the Pythagoras Theorem in the triangles $\triangle ABO$, $\triangle AOA'$ and $\triangle OB'A'$,

$$AA'^2 = AO^2 + OA'^2 = (bc)^2 + (ac)^2 + (bd)^2 + (ad)^2$$
$$= c^2(a^2 + b^2) + d^2(a^2 + b^2) = (a^2 + b^2)(c^2 + d^2).$$

Hence $AA' = \sqrt{a^2 + b^2}\sqrt{c^2 + d^2}$. But $AB \parallel A'B'$ (both being perpendicular to BB'). So AA' is at least as much as the distance BB' between the parallel lines AB and $A'B'$, that is,

$$\sqrt{a^2 + b^2}\sqrt{c^2 + d^2} = AA' \geq BB' = ac + bd.$$

This finishes the proof of the Cauchy-Schwarz inequality.

Equality. If equality holds, then AA' is the distance between the parallel lines AB and $A'B'$ so that AA' is perpendicular to AB and $A'B'$. Thus $AA'B'B$ is a rectangle. Hence the opposite sides AB and $A'B'$ are equal, that is, $bc = ad$, and so $a/c = b/d$.

Vice versa, if $a/c = b/d$, then $AB = bc = ad = A'B'$. Thus again $AA'B'B$ is seen to be a rectangle, now giving the equality $AA' = BB'$. So one has equality in the Cauchy-Schwarz Inequality.

Solution to Exercise 4.17

Let the given point be P in $\triangle ABC$. First let us consider the case when P lies on the side AB, as shown. We have $PR' \parallel AC$ and $PQ \parallel BC$. Let $A_2 := \triangle APQ$, and $A_1 := \triangle PBR'$.

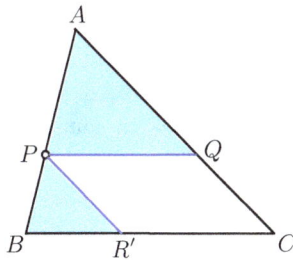

As $PQ \parallel BC$, $\triangle APQ \sim \triangle ABC$, and so $\dfrac{A_1}{A} = \left(\dfrac{AP}{AB}\right)^2$.

As $PR' \parallel AC$, $\triangle PBR' \sim \triangle ABC$, and so $\dfrac{A_2}{A} = \left(\dfrac{PB}{AB}\right)^2$.

Hence $\sqrt{A_1} + \sqrt{A_2} = \dfrac{AP}{AB}\sqrt{A} + \dfrac{PB}{AB}\sqrt{A} = \dfrac{AP + PB}{AB}\sqrt{A} = \dfrac{\cancel{AB}}{\cancel{AB}}\sqrt{A} = \sqrt{A}.$

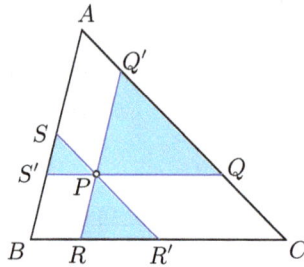

Let $A_1 := \triangle PSS'$, $A_2 := \triangle PQQ'$, $A_3 := \triangle PRR'$. First look at $\triangle AS'Q$. By the previous case, $\sqrt{\triangle AS'Q} = \sqrt{A_1} + \sqrt{A_2}$. If we now "slide" $\triangle PRR'$ along BC, then since $BRPS'$ is a parallelogram, we get the picture below, with $S'R'' \parallel AC$, and $\triangle BS'R'' \simeq \triangle RPR'$.

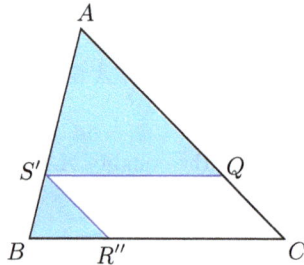

So again by the previous case, we have

$$\sqrt{A} = \sqrt{\triangle AS'Q} + \sqrt{\triangle BS'R''} = \sqrt{\triangle AS'Q} + \sqrt{\triangle RPR'} = \sqrt{\triangle AS'Q} + \sqrt{A_3}.$$

Now using $\sqrt{\triangle AS'Q} = \sqrt{A_1} + \sqrt{A_2}$ (derived earlier), we obtain finally that $\sqrt{A} = \sqrt{A_1} + \sqrt{A_2} + \sqrt{A_3}$.

Solution to Exercise 4.18

The formula $r = (2\Delta)/(a+b+c)$ for the inradius of a triangle $\triangle ABC$ with area Δ and side lengths a, b, c (see Exercise 3.16) shows that the inradius scales in proportion to the sides.

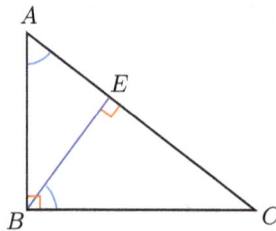

As $\triangle AEB \sim \triangle ABC$, it follows that if r_1 is the inradius of $\triangle AEB$, then

$$r_1 = r\frac{AB}{AC}.$$

Similarly by looking at the similar triangles $\triangle BEC$ and $\triangle ABC$, we have

$$r_2 = r\frac{BC}{AC}.$$

Consequently, using Pythagoras's Theorem,

$$r_1^2 + r_2^2 = r^2\frac{AB^2 + BC^2}{AC^2} = r^2\frac{AC^2}{AC^2} = r^2 \cdot 1 = r^2.$$

Remark 6.3. If R_1, R_2 are the *circum*radii of the two little triangles, and R is the circumradius of $\triangle ABC$, then it is also true that $R_1^2 + R_2^2 = R^2$, but this is obvious, since by Corollary 5.2, the circumradius of a right angled triangle is just half the length of the hypotenuse!

Solution to Exercise 4.19

Using the notation from the proof of Theorem 4.10, we know that $ax \geq cm + bn$. Thus, the Arithmetic Mean-Geometric Mean Inequality (Exercise 5.7) yields

$$ax \geq cm + bn \geq 2\sqrt{bcmn}.$$

Similarly,

$$by \geq c\ell + an \geq 2\sqrt{can\ell},$$
$$cz \geq b\ell + am \geq 2\sqrt{ab\ell m}.$$

Multiplying all these inequalities, we obtain

$$abcxyz \geq 8\sqrt{a^2b^2c^2\ell^2m^2n^2} = 8abc\ell mn.$$

Cancelling $abc > 0$ on both sides gives $xyz \geq 8\ell mn$, that is,

$$PA \cdot PB \cdot PC \geq 8PL \cdot PM \cdot PN.$$

Solutions to the exercises from Chapter 5

Solution to Exercise 5.1

Suppose that the two circles are $C(O, r)$ and $C(O', r')$, intersecting in the points A, B. In triangles $\triangle OAO'$ and $\triangle OBO'$, $OA = OB = r$, $O'A = O'B = r'$ and the sides OO' is common. So by the SSS Congruency Rule, $\triangle OAO' \simeq \triangle OBO'$, giving $\angle AOO' = \angle BOO'$ (CPCT). Denote by M the intersection point of OO' and AB. In the triangles $\triangle OAM$ and $\triangle OBM$, we have $OA = OB = r$, $\angle AOM = \angle BOM$, and the side OM is common. Hence by the SAS Congruency Rule, $\triangle OAM \simeq \triangle OBM$. So $BM = MA$, that is M is the midpoint of AB. Furthermore, $\angle OMA = \angle OMB$, and being supplementary, each equals $90°$. Thus OO' is the perpendicular bisector of AB.

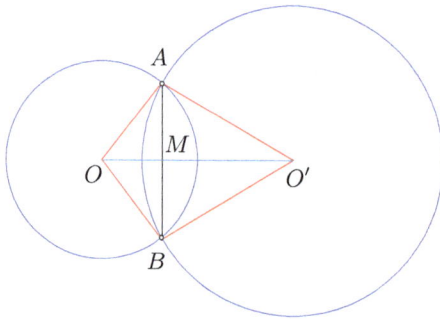

Solution to Exercise 5.2

Let the endpoints of the arc be A, C. Take any point B on the arc between A and C. Draw the perpendicular bisectors of AB, BC. (It can be seen that these perpendicular bisectors aren't parallel because A, B, C are not collinear.) Let O be their point of intersection. Then O is the center of the circle and the radius is OA. So we can complete the circle which contains the given arc.

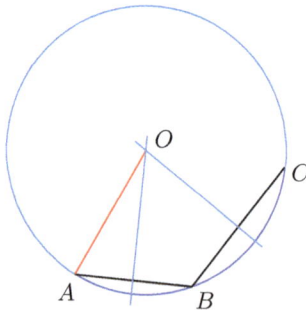

Solution to Exercise 5.3

Consider the line ℓ' which is the common diameter of the big circle and the two semicircles. The line ℓ making a $45°$ angle with ℓ' is the sought-after line. See the picture below. Let A be the area of the big circle, and let A_1, A_2, A_3, A_4, A_5 be the areas of the five regions shown below. Then we have $A_1 = A_5 = A/8$ and $A_2 = A_3$. So we will be done if we manage to show that $A_4 = A_5$. But $A_4 + A_5 = A/4$, and since $A_5 = A/8$, it follows that $A_4 = A/8$ too. So the line ℓ certainly divides the shaded region into two equal parts. A $180°$ rotation of the picture about the center of the big circle, reveals that the line ℓ bisects also the unshaded region!

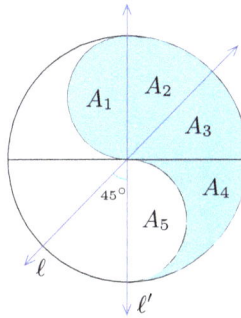

Solution to Exercise 5.4

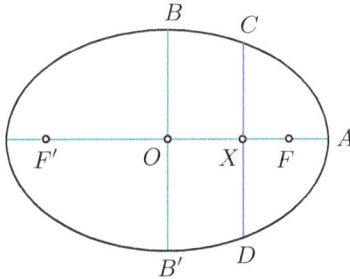

(1) Let $FF' =: d$. Then as $F'B + BF = F'A + FA$, we obtain that

$$2\sqrt{\left(\frac{d}{2}\right)^2 + b^2} = 2a,$$

giving $d = 2\sqrt{a^2 - b^2}$. Next, if $CX =: h$, then as $F'C + CF = F'A + FA$,

$$\sqrt{h^2 + \left(\frac{d}{2} - x\right)^2} + \sqrt{h^2 + \left(\frac{d}{2} + x\right)^2} = 2a.$$

Using the fact that $d = 2\sqrt{a^2 - b^2}$, we can solve for h in the above equation in terms of x, b, a. A bit of messy algebra (eventually) gives $h = (b/a) \cdot \sqrt{a^2 - x^2}$.

(2) By dividing the circle and the ellipse into thin vertical strips, we can add these to obtain the area of the circle and the ellipse.

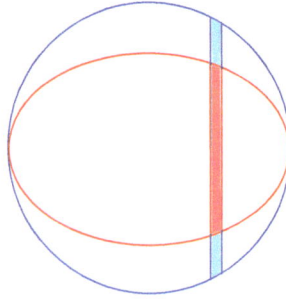

But we note that the length of the chord of the circle at a distance x from the center is $2\sqrt{a^2 - x^2}$, while that of the corresponding chord to the ellipse is $(2b/a) \cdot \sqrt{a^2 - x^2}$. Hence the ratio of lengths of the two strips is b/a. Consequently when added up, we have that the area of the ellipse will be b/a times the area of the circle, that is, $(b/a) \cdot \pi a^2 = \pi ab$.

When b approaches a, the above expression approaches πa^2. (This is expected, since the distance $d = 2\sqrt{a^2 - b^2}$ between the two foci then reduces to 0, and the ellipse becomes the circumscribing circle!)

Solution to Exercise 5.5

By succesive reflection we obtain the following picture. The position of points P, Q determine the positions of the rest of the points P', Q', P'', Q''.

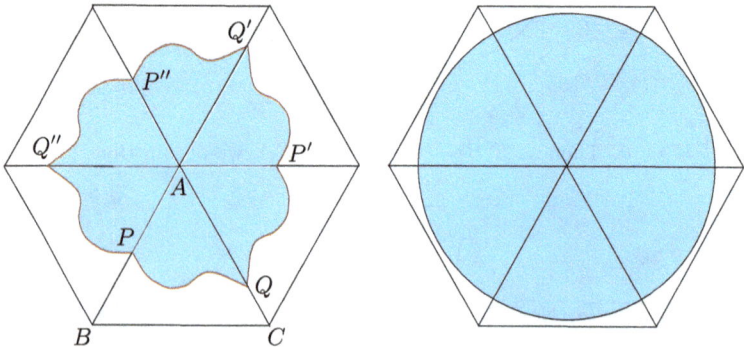

The total length of the curve is fixed, being $L := 6 \cdot \ell$, while the enclosed area by the curve $PQP'Q'P''Q''P$ is 6 times the shaded area in the given problem. It follows from the Isoperimetric Inequality that the area enclosed by the closed curve is maximized for a circle. But then it follows that P, Q should both be equidistant from A, at a distance equal to the radius r of the circle with circumference L, that is,

$$r = AP = AQ = \frac{6 \cdot \ell}{2\pi} = \frac{3\ell}{\pi}.$$

The maximal area is then $\dfrac{\pi r^2}{6} = \dfrac{3\ell^2}{2\pi}.$

Solution to Exercise 5.6

(1) Yes, by the Internal Angle Bisector Theorem, $r = \dfrac{PA}{PB} = \dfrac{AP_1}{BP_1}.$

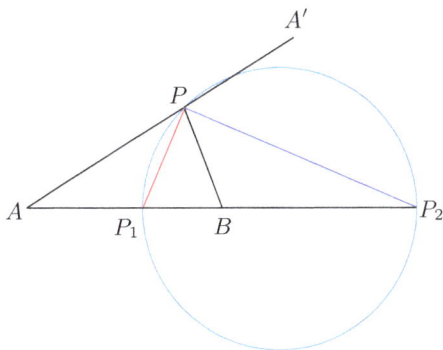

(2) Yes, by the External Angle Bisector Theorem, $r = \dfrac{PA}{PB} = \dfrac{AP_2}{BP_2}.$

(3) $\angle P_1 P P_2 = \angle P_1 PB + \angle BPP_2 = \dfrac{\angle APB}{2} + \dfrac{\angle BPA'}{2} = \dfrac{\angle APA'}{2} = 90°.$

(4) (In the above, the fact that $P_1 P P_2$ is a right angle implies that P lies on the circle with diameter $P_1 P_2$. This motivates the following.)

Let P_1 be the point dividing the line segment AB in the ratio $r > 1$ and let P_2 be the point on AB extended so that $AP_2 : BP_2 = r$. Consider the circle with diameter $P_1 P_2$. This circle, called the Apollonius Circle, is precisely the locus of P. Let us justify this.

First, let P be any point on the locus. Then $PA : PB = r = AP_1 : BP_1$, and so by the converse of the Angle Bisector Theorem, PP_1 is the angle bisector of the internal angle at P in $\triangle APB$. Similarly, $PA : PB = r = AP_2 : BP_2$, and so by the converse of the Exterior Angle Bisector Theorem, PP_2 is the angle bisector of the exterior angle at P in $\triangle APB$. Hence, as shown above, $\angle P_1 PP_2 = 90°$. Consequently, P lies on the circle with diameter $P_1 P_2$.

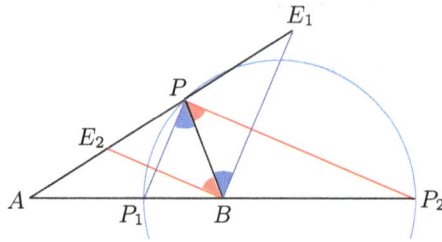

Conversely, suppose that P is any point on the Apollonius Circle. Draw BE_1 parallel to PP_1, meeting AP extended in E_1, and draw BE_2 parallel to PP_2, meeting AP in E_2. Then $\triangle APP_1 \sim \triangle AE_1B$ and $\triangle APP_2 \sim \triangle AE_2B$, so that

$$\frac{AP}{PE_1} = \frac{AP_1}{P_1B} = r = \frac{AP_2}{BP_2} = \frac{AP}{PE_2}. \qquad (6.21)$$

From the equality of the first and the last expressions, we obtain that $PE_1 = PE_2$, so that P is the midpoint of E_1E_2. But from the facts that $\angle P_1PP_2 = 90°$, $P_1P \parallel BE_1$ and $P_2P \parallel BE_2$, it follows that $\angle E_2BE_1 = 90°$ too. If we look at the circle $C(P, PE_1)$ with diameter E_1E_2, then P is the center of $C(P, PE_1)$, and the point B lies on $C(P, PE_1)$. Consequently, $PE_1 = PE_2 = PB$, all being equal to the radius of $C(P, PE_1)$. Now, from (6.21),

$$\frac{AP}{PB} = \frac{AP}{PE_1} = \frac{AP_1}{P_1B} = r,$$

and so P lies on the locus.

(5) If $r > 1$, then depending on the value of r, we get a family of circles which intersect AB on the *right* of the midpoint M of AB. On the other hand, if $r < 1$, we get a family of circles that intersect AB on the left of the midpoint M of AB. If $r = 1$, then the locus is the perpendicular bisector of AB.

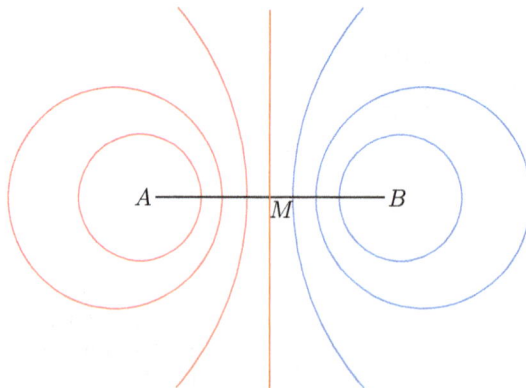

Solution to Exercise 5.7

(1) The radius OO' is half the diameter $AA' = AB + BA' = a + b$.

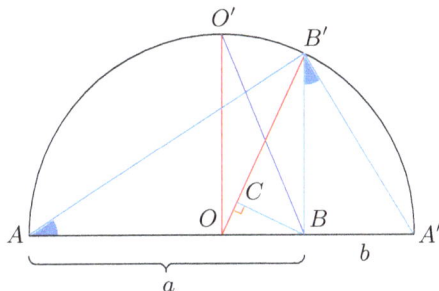

(2) $\angle B'AB = 90° - \angle B'A'B = 90° - (90° - \angle A'B'B) = \angle A'B'B$.

Also $\angle ABB' = 90° = \angle B'BA'$. By the AA Similarity Rule, $\triangle ABB' \sim \triangle B'BA'$.

Thus, $\dfrac{a}{BB'} = \dfrac{BA}{BB'} = \dfrac{BB'}{BA'} = \dfrac{BB'}{b}$, and so $BB'^2 = ab$, that is, $BB' = \sqrt{ab}$.

(3) It is clear that $OB = OA' - BA' = \dfrac{a+b}{2} - b = \dfrac{a-b}{2}$. Thus

$$O'B^2 = O'O^2 + OB^2 = \left(\frac{a+b}{2}\right)^2 + \left(\frac{a-b}{2}\right)^2 = \frac{a^2+b^2}{2}.$$

So $O'B = \sqrt{\dfrac{a^2+b^2}{2}}$.

(4) In the right angled triangles $\triangle OBB'$ and $\triangle BCB'$, we have $\angle OB'B = \angle BB'C$ (common angle), and so $\triangle OBB' \sim \triangle BCB'$. Hence

$$\frac{B'C}{BB'} = \frac{BB'}{B'O},$$

that is, $B'C = \dfrac{BB'^2}{B'O}$. We know $BB'^2 = ab$ and $OB' = \dfrac{a+b}{2}$, the radius of the semicircle. Thus

$$B'C = \frac{ab}{(a+b)/2} = \frac{2}{\dfrac{1}{a} + \dfrac{1}{b}}$$

as claimed.

(5) $\text{QM}_{ab} = O'B > OO' = \text{AM}_{ab} = OB' > BB' = \text{GM}_{ab} > B'C = \text{HM}_{ab}$, where each inequality follows from the fact that the hypotenuse in a right angled triangle is bigger than each of the other sides.

Solution to Exercise 5.8

Let $x =: QP$ and $y =: PR$. We have $\triangle APQ \sim \triangle ACD$ and $\triangle CPR \sim \triangle CAB$.

So $\dfrac{x}{b} = \dfrac{AP}{AC} = \dfrac{AC - PC}{AC} = 1 - \dfrac{PC}{AC} = 1 - \dfrac{PR}{AB} = 1 - \dfrac{y}{a}$, that is, $\dfrac{x}{b} + \dfrac{y}{a} = 1$.

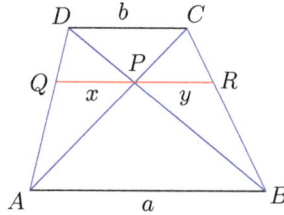

Also, $\triangle DQP \sim \triangle DAB$ and $\triangle BPR \sim \triangle BDC$, and so

$$\frac{x}{a} = \frac{PD}{BD} = \frac{BD - PB}{BD} = 1 - \frac{PB}{BD} = 1 - \frac{PR}{CD} = 1 - \frac{y}{b},$$

that is, $\dfrac{x}{a} + \dfrac{y}{b} = 1$. Adding to this $\dfrac{x}{b} + \dfrac{y}{a} = 1$ (obtained earlier), yields

$$x\left(\frac{1}{a} + \frac{1}{b}\right) + y\left(\frac{1}{a} + \frac{1}{b}\right) = 2,$$

that is, $QR = x + y = \dfrac{2}{1/a + 1/b}$, the harmonic mean of a and b.

Solution to Exercise 5.9

The locus is the interior of the circle with diameter BC minus the diameter BC itself.

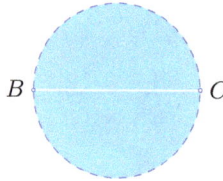

Let I be a point on or outside the semicircle (shown in the leftmost picture of the following figure). If I is the incenter of $\triangle ABC$, we have

$$\angle ABC + \angle ACB = 2\angle IBC + 2\angle ICB$$
$$\geq 2\angle PBC + 2\angle PCB = 2(180° - \angle BPC)$$
$$= 2(180° - 90°) = 2 \cdot 90° = 180°,$$

a contradiction.

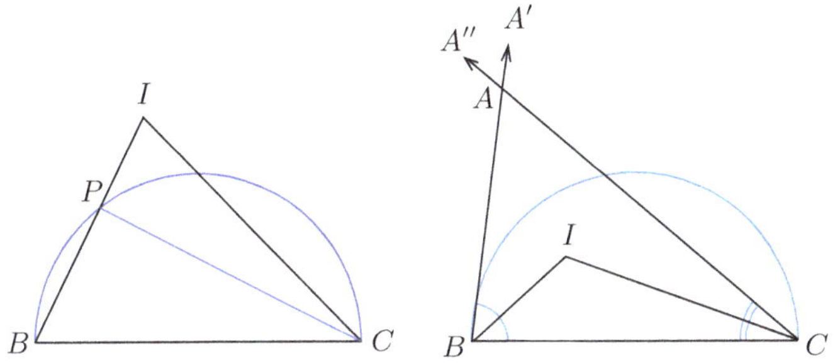

On the other hand, if I lies inside the semicircle as shown in the rightmost picture above, and I does not lie on BC, then let us show that there exists a point A such that I is the incenter of $\triangle ABC$. Join I to B and to C. We can draw a ray $\overrightarrow{BA'}$ so that $\angle A'BC = 2\angle IBC$. Similarly, we can draw a ray $\overrightarrow{CA''}$ so that $\angle A''CB = 2\angle ICB$. Then

$$\angle A'BC + \angle A''CB = 2\angle IBC + 2\angle ICB = 2(180° - \angle BIC) < 2(180° - 90°) = 180°.$$

So with BC as transversal to the lines $\overleftrightarrow{BA'}$ and $\overleftrightarrow{CA''}$, it follows that since the corresponding interior angles on one side of BC have a sum strictly less than $180°$, these lines intersect on that side of BC. If A is the intersection point of these lines, then we see that I is the incenter of $\triangle ABC$.

Solution to Exercise 5.10

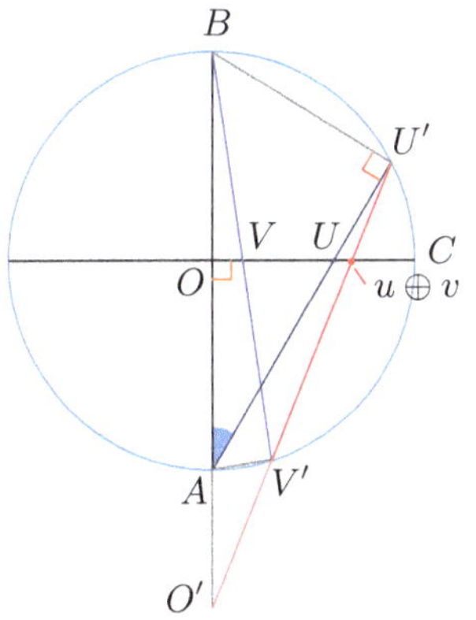

As $\angle AOU = 90° = \angle AU'B$ and $\angle OAU = \angle U'AB$ (common), by the AA Similarity Rule, $\triangle AOU \sim \triangle AU'B$. So

$$\frac{AU'}{2} = \frac{AU'}{AB} = \frac{AO}{AU} = \frac{1}{\sqrt{1+u^2}},$$

giving $AU' = 2/\sqrt{1+u^2}$. Hence

$$UU' = AU' - AU = 2/\sqrt{1+u^2} - \sqrt{1+u^2} = (1-u^2)/\sqrt{1+u^2}.$$

Proceeding similarly, we obtain $BV' = 2/\sqrt{1+v^2}$ and $VV' = (1-v^2)/\sqrt{1+v^2}$. Let W be the point of intersection of $U'V'$ and OC, and set $OW =: w$. Menelaus's Theorem applied to $\triangle AOU$ with the line $O'U'$ gives

$$\frac{w-u}{w} \cdot \frac{OO'}{OO'-1} \cdot \frac{2/\sqrt{1+u^2}}{(1-u^2)/\sqrt{1+u^2}} = \frac{UW}{OW} \cdot \frac{OO'}{AO'} \cdot \frac{AU'}{UU'} = 1.$$

This yields

$$\frac{1}{OO'} = 1 - \frac{2}{1-u^2} \cdot \frac{w-u}{w}. \tag{6.22}$$

Similarly, Menelaus's Theorem applied to $\triangle BOV$ with the line $O'U'$ gives

$$\frac{w-v}{w} \cdot \frac{OO'}{OO'+1} \cdot \frac{2/\sqrt{1+v^2}}{(1-v^2)/\sqrt{1+v^2}} = \frac{VW}{OW} \cdot \frac{OO'}{BO'} \cdot \frac{BV'}{VV'} = 1.$$

This yields

$$\frac{1}{OO'} = \frac{2}{1-v^2} \cdot \frac{w-v}{w} - 1. \tag{6.23}$$

Equating the right-hand sides of (6.22) and (6.23) gives, after some algebraic manipulations, that

$$w = \frac{u+v}{1+uv} = u \oplus v.$$

If $u, v \ll 1$, then $\angle OBV \approx 0$, and AV' is almost parallel to OV. So $\triangle BOV$ is almost similar to $\triangle BAV'$, giving

$$AV' \approx \frac{AB}{OB} \cdot OV = 2 \cdot OV = 2v.$$

Since AV' is almost parallel to OC, $\triangle U'UW$ is almost similar to $\triangle U'AV'$. Moreover, as $u, v \ll 1$, $U'V' \approx AB = 2$, and $U'W \approx OB = 1$. Hence

$$UW \approx \frac{U'W}{U'V'} \cdot AV' \approx \frac{1}{2} \cdot 2v = v.$$

Thus $w - u = UW \approx v$, that is, $w \approx u + v$.

Solution to Exercise 5.11

In the following picture, O, O' are the centers of the two smaller circles, and P, P' are the points of contact of these circles with the big circle C.

From the picture, it is clear that the other points X on the original circle which are on the same side of AB as the point P give smaller angles as they are outside the internally touching small circle (with center O).

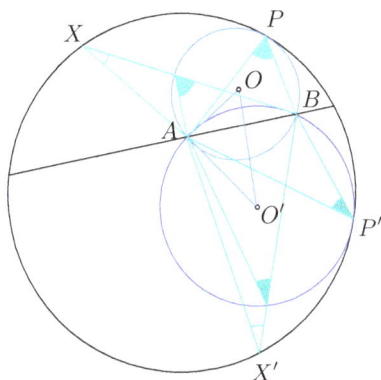

The angle at the point of contact of the bigger internally touching circle (with center O') is also smaller than the angle subtended by AB at P, and here's why. Indeed, in $\triangle AOO'$, $\angle AOO' \geq \angle AO'O$ since $AO' \geq AO$. So using the fact that $\triangle AOO' \simeq \triangle BOO'$ by the SSS Rule of Congruency, and using Theorem 5.6,

$$\angle APB = \frac{\angle AOB}{2} = \angle AOO' \geq \angle AO'O = \frac{\angle AO'B}{2} = \angle AP'B.$$

So at any point X' on the original circle $\mathcal{C}$ which lies on the same side of AB as the point P', we see that AB will subtend a smaller angle (than the angle subtended at P), since X' is outside the larger touching circle with center O', giving $\angle AX'B < \angle AP'B \leq \angle APB$.

Solution to Exercise 5.12

Consider the circumcircle $\mathcal{C}$ of $\triangle ABP$. If Q lies on the $\mathcal{C}$, then we are done. If not, then join A to Q, and let it (possibly extended) meet $\mathcal{C}$ at Q'.

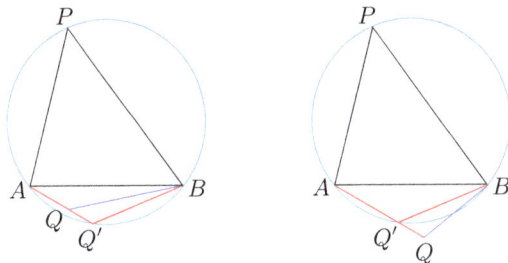

$\angle AQ'B = 180° - \angle APB = \angle AQB$. On the other hand, by looking at the triangle $\triangle BQQ'$, we see that either $\angle AQB > \angle AQ'B$ (left picture above), or $\angle AQ'B < \angle AQB$ (right picture above). In either case we have a contradiction.

Solution to Exercise 5.13

Since $r > 1$, we have $r^2 > r > 1$, and so the side r^2 must be the hypotenuse. By Pythagoras's Theorem,

$$(r^2)^2 = r^2 + 1^2.$$

As $r^2 > 0$, we discard the negative root of the above quadratic expression in r^2, and so

$$r^2 = \frac{1 + \sqrt{5}}{2} = \tau,$$

that is, $r = \sqrt{\tau}$.

Solution to Exercise 5.14

Let $\tau \in \mathbb{Q}$. Then some scaling of the golden rectangle with side lengths 1 and τ will produce a golden rectangle with integer sides.

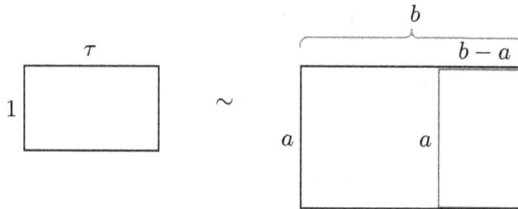

Among all such similar golden rectangles with integer side lengths, arrange them in increasing order by longest side length. Consider the first rectangle R in this list. Then R a golden rectangle with integer side lengths, say a, b, in which the longest side b ($= \tau a > a$) is the smallest possible. But if we cut out a square with side length equal to a, then we are left with a smaller golden rectangle, with integer sides, and with longest side a ($< b$), contradicting our choice of R. Thus τ can't be rational. Hence $\sqrt{5} = 2\tau - 1$ can't be rational either.

Solution to Exercise 5.15

(1) Dividing the recurrence rule throughout by F_n, we obtain $\dfrac{F_{n+1}}{F_n} = 1 + \dfrac{F_{n-1}}{F_n}$.

For large n, $\dfrac{F_{n+1}}{F_n}, \dfrac{F_{n-1}}{F_n} \approx L > 0$, and so from the above, we obtain $L = 1 + \dfrac{1}{L}$.

As $L > 0$, we discard the negative root of the quadratic, giving $L = \dfrac{1 + \sqrt{5}}{2} = \tau$.

(2) One mile is about 1.609 kilometers, and this is near $\tau \approx 1.618$, which in turn is near F_{n+1}/F_n for large n.

(3) The areas of the little squares add up to the area of the big rectangle.

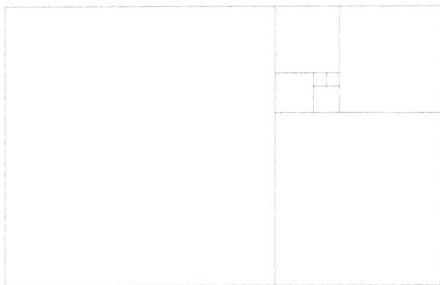

So $F_0^2 + F_1^2 + \cdots + F_n^2 = F_n \cdot F_{n+1}$.

Solution to Exercise 5.16

Let the right angled triangle be $\triangle ABC$, right angled at B. Complete it to a rectangle $ABCD$. Then the points A, B, C, D are concyclic (for example because the angles subtended by BC at A and at D are equal, which in turn follows from the fact that $\triangle ABC \simeq \triangle DCB$).

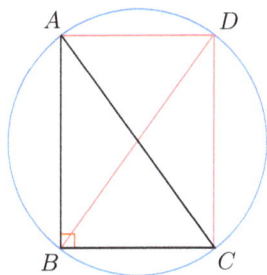

By Ptolemy's Theorem, and by using the fact that $ABCD$ is a rectangle (and so opposite pairs of sides are equal, and so are the two diagonals), we obtain

$$
\begin{aligned}
AC^2 &= AC \cdot AC = AC \cdot BD \\
&= AB \cdot CD + AD \cdot BC \\
&= AB \cdot AB + BC \cdot BC \\
&= AB^2 + BC^2;
\end{aligned}
$$

Pythagoras's Theorem!

Solution to Exercise 5.17

Without loss of generality, we can assume that P lies on the arc between B and C. Referring to the picture below, we have, by Ptolemy's Theorem applied to the cyclic quadrilateral $ABPC$ that $AB \cdot PC + CA \cdot PB = BC \cdot PA$.

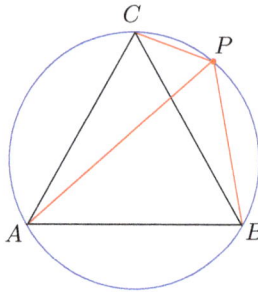

But as $AB = BC = CA$, it follows from here that $PC + PB = PA$.

Solution to Exercise 5.18

Let DE extended meet the circle in F, F'.

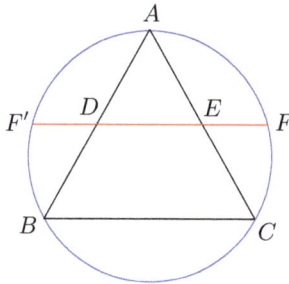

Let $a := AB$. Then we have, by Theorem 5.14, that

$$DF' \cdot (DE + EF) = DF' \cdot DF = AD \cdot DB = \frac{a^2}{4}$$
$$= AE \cdot EC = F'E \cdot EF = (DE + DF') \cdot EF.$$

From the equality of the first and last expression, we obtain $DE \cdot (EF - DF') = 0$, and since $DE \neq 0$, it follows that $EF = DF'$. By the Midpoint Theorem, $DE = a/2$, and so if $DE : EF = k$, then we have, using $AE \cdot EC = F'E \cdot EF$, that

$$k^2 EF^2 = DE^2 = \frac{a^2}{4} = AE \cdot EC = F'E \cdot EF$$
$$= (DE + EF) \cdot EF = (kEF + EF) \cdot EF,$$

and so $k^2 = k + 1$. As $k > 0$, it follows that $k = (1 + \sqrt{5})/2 = \tau$.

Solution to Exercise 5.19

Let $d = OI$ denote the distance between the circumcenter O and the incenter I of $\triangle ABC$. Then from Euler's Theorem, $d^2 = R^2 - 2Rr = R(R - 2r)$.

(1) As $d^2 \geq 0$, and since $R > 0$, it follows that $R - 2r = d^2/R \geq 0$, so that $R \geq 2r$.

(2) If $R = 2r$, then $d = 0$, and so the incenter and the circumcenter coincide. But then in the two isosceles triangles $\triangle ABO$ and $\triangle ACO$, $AO = OB = OC$, and as $AO = AI$ is the angle bisector of $\angle A$, also $\angle BAO = \angle CAO$. So these two triangles are congruent. Hence $AB = AC$. Similarly, $AB = AC$ as well. Thus $\triangle ABC$ is equilateral.

On the other hand, if $\triangle ABC$ is equilateral, then it is clear that perpendicular bisectors of sides bisect the angle opposite to the side. Hence the point of concurrency of the perpendicular bisectors (the circumcenter O), must coincide with the point of concurrency of the angle bisectors (namely the incenter I). Hence $d = OI = 0$, so that $R = 2r$.

(3) Write $R = 2r + \delta$. Then note by (1) that $\delta \geq 0$.

We have $\dfrac{R}{r} + \dfrac{r}{R} - \dfrac{5}{2} = \dfrac{2r + \delta}{r} + \dfrac{r}{2r + \delta} - \dfrac{5}{2} = \dfrac{3r\delta + 2\delta^2}{2r(2r + \delta)} \geq 0.$

Equality holds if and only if $\dfrac{3r\delta + 2\delta^2}{2r(2r + \delta)} = 0$, that is, if and only if $\delta = 0$.

But $\delta = 0$ if and only if $R = 2r$, and so by part (2), we have equality if and only if the triangle is equilateral.

Solution to Exercise 5.20

See the pictures below, where we have dropped a perpendicular from A to the side BC, drawn the diameter AE through A, and joined B to E.

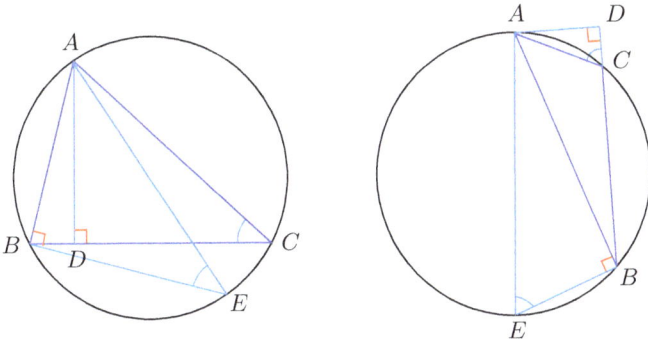

In triangles $\triangle ADC$ and $\triangle ABE$, we have that $\angle ADC = 90° = \angle ABE$. The latter equality holds since the diameter of a circle subtends a right angle at any point of the semicircle described by it.

Also, if the points C, E lie on the same arc of the chord AB (that is, either on the major arc, or on the minor arc; see the left picture above), then since the chord AB subtends equal angles at the points C and E, we have that $\angle AEB = \angle ACD$.

If, on the other hand, one among C, E lies on the major arc, and the other on the minor arc, then using the supplementarity of the angles subtended by a chord in the major/minor arc, we again obtain (see the rightmost picture), that $\angle ACD = 180° - \angle ACB = 180° - (180° - \angle AEB) = \angle AEB$. By the AA Similarity Rule, $\triangle ADC \sim \triangle ABE$.

Thus $\dfrac{AD}{AB} = \dfrac{AC}{AE}$. So $AD = \dfrac{AC \cdot AB}{AE} = \dfrac{bc}{2R}$.

Consequently, $\Delta = \dfrac{BC \cdot AD}{2} = \dfrac{abc}{2 \cdot 2R}$, that is, $R = \dfrac{abc}{4\Delta}$.

Remark 6.4. Note that in this formula, if we use Heron's Formula for the area Δ, we get the following expression for the circumradius entirely in terms of the side lengths a, b, c:

$$R = \frac{abc}{\sqrt{(a+b+c)(b+c-a)(c+a-b)(a+b-c)}}.$$

Solution to Exercise 5.21

If a, b, c denote the lengths of the sides of the triangle, then

$$2\Delta = 2\Delta ABC = 2(\Delta PBC + \Delta PCA + \Delta PAB) = ax + by + cz.$$

By the Arithmetic Mean-Geometric Mean Inequality,

$$\frac{2\Delta}{3} = \frac{ax + by + cz}{3} \geq \sqrt[3]{ax \cdot by \cdot cz},$$

and so using the result in Exercise 5.20, we obtain $xyz \leq \dfrac{8\Delta^3}{27abc} = \dfrac{2\Delta^2}{27R}$.

Equality holds if and only if $ax = by = cz$, that is, $\Delta PBC = \Delta PCA = \Delta PAB$.

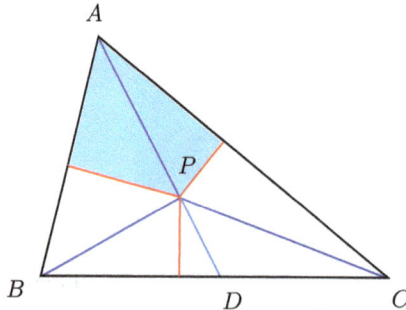

Let us show the "if part". Join A to P and extend it to meet BC in D. Then if H, h are the lengths of the altitudes dropped from A, P to BC, respectively, then we have

$$\frac{BD \cdot H}{2} - \frac{BD \cdot h}{2} = \triangle ABD - \triangle PBD = \triangle PAB = \triangle PCA = \triangle ADC - \triangle PDC$$
$$= \frac{DC \cdot H}{2} - \frac{DC \cdot h}{2},$$

and so $BD = DC$. Similarly, we can show that BP extended meets AC in the midpoint of AC, and so P is indeed the centroid.

The "only if part" is straightforward.

Solution to Exercise 5.22

(1) Let the three circles be labelled as $1, 2, 3$. Suppose that circles $1, 2$ intersect at A_{12}, A'_{12}, circles $2, 3$ intersect at A_{23}, A'_{23}, and circles $3, 1$ intersect at A_{31}, A'_{31}. Suppose that $A_{12} A'_{12}$ and $A_{23} A'_{23}$ intersect at the point P. Join A_{31} to P, and suppose that it meets the circles $1, 2$ at points B_1, B_2, respectively.

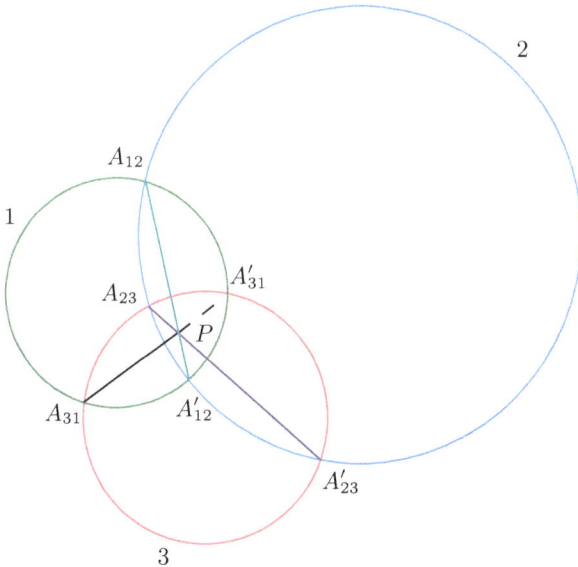

By Theorem 5.14, we obtain

$$A_{12} P \cdot P A'_{12} = A_{31} P \cdot P B_1,$$
$$A_{12} P \cdot P A'_{12} = A_{23} P \cdot P A'_{23},$$
$$A_{31} P \cdot P B_2 = A_{23} P \cdot P A'_{23}.$$

Hence $A_{31}P \cdot PB_2 = A_{31}P \cdot PB_1$, giving $PB_2 = PB_1$, so that $B_2 = B_1$. Consequently $B_2 = B_1$ is a point on both circles $1,3$, and so it must be A'_{31}. This completes the proof.

Remark 6.5. There are other elegant proofs of this beautiful result. A proof based on Cartesian coordinate geometry, attributed to Plücker, is as follows. Suppose that the three circles are given by $f_1(x,y) = 0$, $f_2(x,y) = 0$, $f_3(x,y) = 0$, where each function f_i conatins $x^2 + y^2$ plus linear terms $a_ix + b_iy + c_i$. By subtracting these equations in pairs, we note that the quadratic terms cancel, leaving the equations of the three lines containing the three common chords: $f_1 - f_2 = 0$, $f_2 - f_3 = 0$, $f_3 - f_1 = 0$. But it is clear that any two of these linear equations implies the third left over one, showing that the point of intersection of any two chords also lies on the remaining chord!

The result is also intuitively expected, based on our physical know-how of the three dimensional world, and we elaborate on this now. We will "lift" the problem to three dimensions by considering three hemispherical cups on a horizontal plane, whose boundaries are the three given circles in our (horizontal) plane. Note that any two hemispherical cups intersect in a semicircle, whose "shadow" from a light source placed high above is precisely the common chord for our circles. But the three hemispherical surfaces come together at a point — a dip — where the three semicircles corresponding to pairwise intersections meet. If we imagine rain coming down from above, this "dip" will be where water starts collecting, forming a little puddle. The shadow of this dip on the horizontal plane is the common intersection point of the three pairwise-common chords.

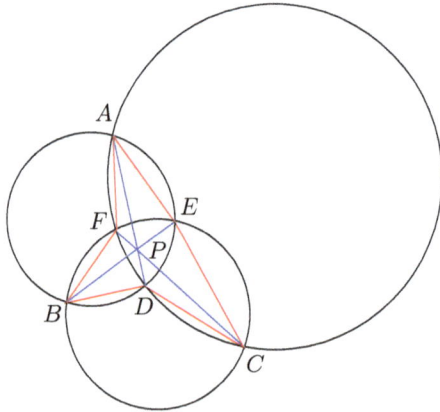

(2) See the picture above. Consider the circle passing through A, E, D, B. Its chords AD, BE intersect at P. Then as in the proof of Theorem 5.14, we have $\triangle AEP \sim \triangle BDP$, and so
$$\frac{BD}{AE} = \frac{DP}{EP}.$$
Similarly, by considering the circle which passes through A, F, D, C, and its chords

AD, CF intersecting at P, we see that $\triangle AFP \sim \triangle CDP$, so that

$$\frac{AF}{DC} = \frac{FP}{DP}.$$

Finally, by considering the circle which passes through B, F, E, C, and its chords BE, CF intersecting at P, we see that $\triangle BFP \sim \triangle CEP$, so that

$$\frac{CE}{FB} = \frac{EP}{FP}.$$

Consequently, $\dfrac{BD}{DC} \cdot \dfrac{CE}{EA} \cdot \dfrac{AF}{FB} = \dfrac{BD}{AE} \cdot \dfrac{AF}{DC} \cdot \dfrac{CE}{FB} = \dfrac{DP}{EP} \cdot \dfrac{FP}{DP} \cdot \dfrac{EP}{FP} = 1.$

Solution to Exercise 5.23

This follows immediately from Exercise 5.12. Indeed, consider the diagonal AC. Then we are given that the points B, D on either side of AC are such that $\angle ABC + \angle ADC = 180°$. By the result established in Exercise 5.12, it follows that A, B, C, D are concyclic, that is $ABCD$ is a cyclic quadrilateral.

Solution to Exercise 5.24

Let $ABCD$ be a cyclic parallelogram. As opposite angles are equal in a parallelogram, we have $\angle A = \angle C$ and $\angle B = \angle D$. On the other hand, thanks to cyclicity, these opposite angles are also supplementary, showing that each must be $90°$. Hence the parallelogram is a rectangle.

Solution to Exercise 5.25

Let $ABCD$ be the cyclic quadrilateral such that $AB \parallel CD$. Extend AD, BC to meet at X. See the picture below.

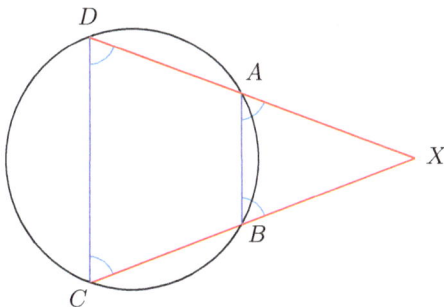

As $AB \parallel CD$, we have $\angle XAB = \angle XDC$ and $\angle XBA = \angle XCD$. Moreover,

$$\angle XAB = 180° - \angle DAB$$
$$= \angle BCD \quad \text{(as } ABCD \text{ is cyclic)}$$
$$= \angle XCD.$$

So combining the above observations, we obtain that

$$\angle XAB = \angle XDC = \angle XCD = \angle XBA.$$

In particular, in $\triangle XAB, \triangle XDC$, $\angle XAB = \angle XBA$ and $\angle XDC = \angle XCD$, and so they are both isosceles with $XA = XB$ and $XD = XC$. Subtracting these last equations, we conclude that $AD = XD - XA = XC - XB = BC$.

Solution to Exercise 5.26

Let $\alpha, \beta, \gamma, \delta$ denote the angles shown in the picture below.

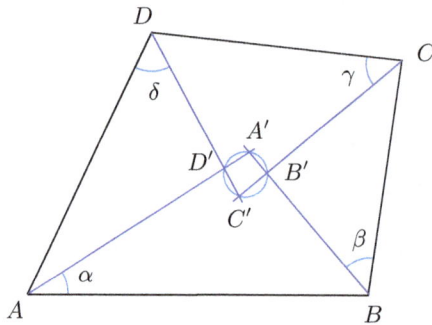

We have $\angle AA'B = 180° - (\alpha + \beta)$ and $\angle DC'C = 180° - (\delta + \gamma)$. Thus

$$\angle AA'B + \angle DC'C = 180° - (\alpha + \beta) + 180° - (\delta + \gamma)$$
$$= 360° - (\alpha + \beta + \gamma + \delta)$$
$$= 360° - \left(\frac{360°}{2}\right) = 180°.$$

Consequently $A'B'C'D'$ is cyclic.

Solution to Exercise 5.27

(1) As $\angle AXP + \angle AYP = 90° + 90° = 180°$, it follows that the quadrilateral $AXPY$ is cyclic. Similarly the quadrilateral $AY'DX'$ is also cyclic.

We note that $\angle XAP = \angle BAD' - \angle D''AD' = \angle D'AC - \angle D'AD = \angle DAX'$. The chord XP subtends equal angles on one side of the circle passing through A, X, P, Y, and so $\angle XYP = \angle XAP$. Similarly, by considering the chord DX' of the circle passing through Y', A, X', D, we obtain that $\angle DY'X' = \angle DAX'$.

Consider the two triangles $\triangle XPY$ and $\triangle X'DY'$. We have

$$\angle XYP = \angle XAP = \angle DAX' = \angle DY'X',$$

and also $\angle XPY = 180° - \angle A = \angle Y'DX'$. By the AA Similarity Rule, we have $\triangle XPY \sim \triangle X'DY'$. Consequently,

$$\frac{x}{y} = \frac{XP}{PY} = \frac{DX'}{DY'} = \frac{AC \cdot DX'}{AB \cdot DY'} \cdot \frac{c}{b} = \frac{2\triangle ADC}{2\triangle ABD} \cdot \frac{c}{b} = \frac{c}{b}.$$

(2) Drop the altitude from A to the side BC. Then $\triangle ABD''$ and $\triangle AD''C$ share this altitude. Using this, and by taking $P = D''$ in the previous part, we have that

$$\frac{BD''}{D''C} = \frac{\triangle ABD''}{\triangle AD''C} = \frac{D''X \cdot AB}{D''Y \cdot AC} = \frac{x}{y} \cdot \frac{c}{b} = \frac{c}{b} \cdot \frac{c}{b} = \frac{c^2}{b^2}.$$

(3) Suppose that the three symmedians are AD'', BE'', CF''. Then we have

$$\frac{BD''}{D''C} \cdot \frac{CE''}{E''A} \cdot \frac{AF''}{F''B} = \frac{c^2}{b^2} \cdot \frac{a^2}{c^2} \cdot \frac{b^2}{a^2} = 1.$$

By the converse of Ceva's Theorem, AD'', BE'', CF'' are concurrent.

(4) It can be checked easily that

$$(x^2 + y^2 + z^2)(a^2 + b^2 + c^2) = (xc + yb + za)^2$$
$$+(bx - cy)^2 + (ay - bz)^2 + (cz - ax)^2,$$

so that $(xc + yb + za)^2 \leq (x^2 + y^2 + z^2)(a^2 + b^2 + c^2)$, with equality if and only if $0 = bx - cy = ay - bz = cz - ax$, that is, if and only if

$$\frac{x}{y} = \frac{c}{b} \quad \text{and} \quad \frac{y}{z} = \frac{b}{a} \quad \text{and} \quad \frac{z}{x} = \frac{a}{c}.$$

If P is a point inside the triangle, with distances x, y, z to AB, AC, BC, then

$$2\triangle ABC = 2(\triangle APB + \triangle APC + \triangle BPC)$$
$$= xc + yb + za$$
$$\leq \sqrt{(a^2 + b^2 + c^2)(x^2 + y^2 + z^2)},$$

and so $x^2 + y^2 + z^2$ is minimized if P is the Lemoine Point.

Solution to Exercise 5.28

As d goes to 0, we have that $s = \dfrac{a+b+c+d}{2}$ approaches

$$s' := \frac{a+b+c+0}{2} = \frac{a+b+c}{2},$$

the semiperimeter of a triangle with side lengths a, b, c. The expression for the area of the cyclic quadrilateral, namely $\sqrt{(s-a)(s-b)(s-c)(s-d)}$, thus approaches $\sqrt{(s'-a)(s'-b)(s'-c)(s'-0)} = \sqrt{s'(s'-a)(s'-b)(s'-c)}$, which is exactly Heron's Formula for the area of a triangle with side lengths a, b, c.

   This is expected, since when d diminished to 0, the pair of the vertices corresponding to that side of length d of the cyclic quadrilateral merge into one vertex, and the cyclic quadrilateral degenerates to a triangle with side lengths a, b, c.

Solution to Exercise 5.29

Let the tangency points be denoted by M_a, M_b, M_c, M_d. Then we have by Theorem 5.23 that $AM_a = AM_d =: x$, $BM_b = BM_a =: y$, $CM_c = CM_b =: z$, $DM_d = DM_c =: w$, and so $a = x+y$, $b = y+z$, $c = z+w$, $d = w+x$.

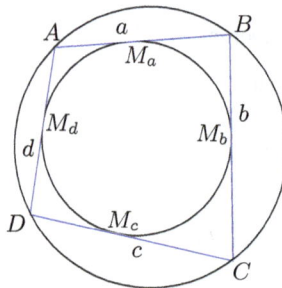

Thus $s = (a+b+c+d)/2 = x+y+z+w$, and

$$
\begin{aligned}
s - a &= x+y+z+w - (x+y) = z+w = c,\\
s - b &= x+y+z+w - (y+z) = x+w = d,\\
s - c &= x+y+z+w - (z+w) = x+y = a,\\
s - d &= x+y+z+w - (w+x) = y+z = b.
\end{aligned}
$$

Hence the area of $ABCD$ is $\sqrt{(s-a)(s-b)(s-c)(s-d)} = \sqrt{abcd}$.

Solution to Exercise 5.30

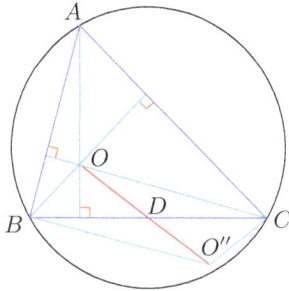

Let O'' be a point such that D is the midpoint of OO''. In the quadrilateral $OBO''C$, the diagonals OO'' and BC bisect each other, and so it follows from Exercise 3.2 that $OBO''C$ has to be a parallelogram. Hence $BO \parallel O''C$. But $\overleftrightarrow{BO}$ is perpendicular to AC, and so $\angle O''CA = 90°$. Similarly, $\angle O''BA = 90°$ too. So the points A, B, O'', C are concyclic. So O'' lies on the extension of OD, as well as on the circumcircle of $\triangle ABC$. Thus $O = O''$, proving the claim that D bisects $OO' = OO''$.

Solution to Exercise 5.31

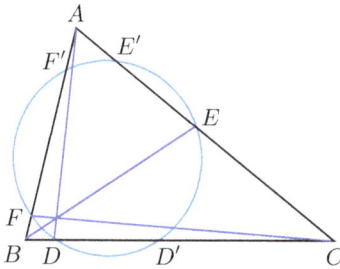

By Ceva's Theorem, $\dfrac{BD}{DC} \cdot \dfrac{CE}{EA} \cdot \dfrac{AF}{FB} = 1$. By Theorem 5.18,

$$BD \cdot BD' = FB \cdot F'B, \quad DC \cdot D'C = CE \cdot CE', \quad EA \cdot E'A = AF \cdot AF'.$$

So $\dfrac{BD'}{D'C} \cdot \dfrac{CE'}{E'A} \cdot \dfrac{AF'}{F'B} = \dfrac{BD'}{F'B} \cdot \dfrac{CE'}{D'C} \cdot \dfrac{AF'}{E'A} = \dfrac{FB}{BD} \cdot \dfrac{DC}{CE} \cdot \dfrac{EA}{AF} = 1.$

By the converse of Ceva's Theorem, AD', BE', CF' are concurrent.

Solution to Exercise 5.32

Suppose that initially the point of contact is $P_0 = C_0$, and we consider the position P_t of this point on the smaller circle at a later time, when the new point of contact is, say C_t. Let the center of the big circle be denoted by O. Let O_t denote the center of the little circle when the point of contact is C_t.

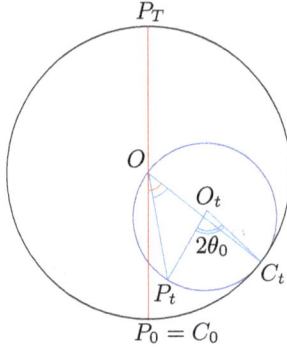

Since the little circle moves without slipping, it follows that the length of the circular arc with endpoints P_0, C_t on the big circle, is the same as that of the length of the circular arc with end points P_t, C_t on the little circle. So if $\angle P_0 O C_t = \theta_0$ degrees, and $\angle P_t O_t C_t = \theta_t$ degrees, then

$$\frac{\theta_t \pi r}{180} = \frac{\theta_0 \pi (2r)}{180}.$$

Consequently, $\angle P_t O_t C_t = \theta_t = 2\theta_0$. But $\angle P_t O C_t$ is the angle subtended by the chord $P_t C_t$ in the smaller circle, and this will be half the angle $\angle P_t O_t C_t$ subtended by $P_t C_t$ at the center O_t. So

$$\angle P_t O C_t = \frac{\angle P_t O_t C_t}{2} = \frac{2\theta_0}{2} = \theta_0 = \angle P_0 O C_t.$$

Hence the rays $\overrightarrow{OP_0}$ and $\overrightarrow{OP_t}$ make the same angle θ_0 on the same side of the ray $\overrightarrow{OC_t}$. This implies that O, P_0, P_t lie on the same line. Hence P_t always lies on the diameter of the big circle passing through P_0. Suppose this diameter meets the big circle at the point P_T diametrically opposite to P_0. Note that when the little circle makes one full rotation, the initial point P_0 on the little circle ends up at the point P_T at some later time T. As the motion is continuous, we see that all the points on the diameter $P_0 P_T$ are covered by the path traced out by P_t as the time t increases. So the path traced out is the diameter $P_0 P_T$.

Solution to Exercise 5.33

Through the point T, draw a chord TS. Let R by any point on the major arc described by TS. Join R to T and to S. Draw a ray $\overrightarrow{TP}$ so that P lies in that side of the line TS which doesn't contain the point R, and such that $\angle TRS = \angle STP$. Then by Theorem 5.22, TP is the required tangent line.

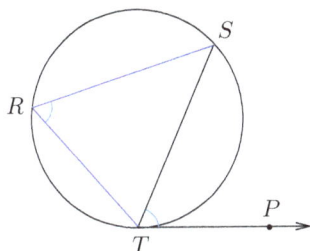

Solution to Exercise 5.34

Draw a line segment BC, and draw a ray $\overrightarrow{BP}$ such that $\angle CBP$ has the given angle measure $\angle A$. Now erect a line ℓ which is perpendicular to BP through the point B. Construct the perpendicular bisector of BC, and suppose that it meets ℓ at O on the side of BC which is opposite to the side containing the point P. With O as center and radius OB, draw a circle $C(O, OB)$. With the midpoint D of BC taken as the center, and with radius equal to the given length of the median AD, draw a circle $C(D, AD)$, and suppose that it intersects the circle $C(O, OB)$ at the points A, A'. Then $\triangle ABC$ (or just as well $\triangle A'BC$) is the required triangle.

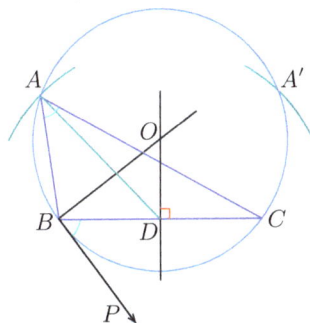

Rationale: As $\angle OBP = 90°$, $\overrightarrow{BP}$ is tangential to the circle $C(O, OB)$ with point of tangency B. Hence $\angle BAC = \angle CBP$, and the latter was ensured to have measure $\angle A$.

Solution to Exercise 5.35

Draw a line segment BC, and draw a ray $\overrightarrow{BP}$ such that $\angle CBP$ has the given angle measure $\angle A$. Now erect a line ℓ which is perpendicular to BP through the point B. Construct the perpendicular bisector of BC, and suppose that it meets the line ℓ at O on the side of BC which is opposite to the side containing the point P. With O as center and radius OB, draw a circle $C(O, OB)$. With the midpoint M of BC taken as the center, and with radius equal to the given length of the altitude AD, draw a circle $C(M, AD)$, and suppose that it intersects the perpendicular bisector of BC at the point L on the same side of BC as the point O. Draw a line through L which is parallel to BC, and suppose that it meets the circle $C(O, OB)$ at the points A, A'. Then $\triangle ABC$ (or just as well $\triangle A'BC$) is the required triangle.

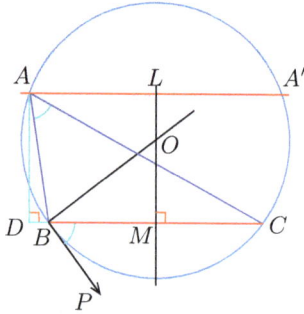

Rationale: As $\angle OBP = 90°$, $\overrightarrow{BP}$ is tangential to the circle $C(O, OB)$ with point of tangency B. Hence $\angle BAC = \angle CBP$. Also, the altitude $AD \parallel LM$ (since both are perpendicular to BC), and $LMDA$ is a rectangle with $LM = AD$.

Solution to Exercise 5.36

Using $R_\oplus = 6400$ km, and the relations

$$\frac{d_{\mathbb{C}}}{d_\odot} \approx \frac{1}{381}, \quad \frac{d_{\mathbb{C}}}{d_\odot} = \frac{R_{\mathbb{C}}}{R_\odot}, \quad \frac{R_{\mathbb{C}}}{R_\oplus} = \frac{T_{\mathbb{C}}}{T_\oplus} = \frac{0.93}{3.39},$$

$$2\pi d_{\mathbb{C}} = (28 \times 24 \text{ hours}) \cdot \left(\frac{2 \times R_{\mathbb{C}}}{T_{\mathbb{C}}} \frac{\text{km}}{\text{hours}} \right),$$

we get $R_{\mathbb{C}} = \dfrac{0.93}{3.39} \cdot 6400 \approx 1756$ km, $d_{\mathbb{C}} = \dfrac{(28 \times 24) \times (2 \times 1756)}{2\pi \cdot 0.93} \approx 404093$ km,

$d_\odot = 381 \times 404093 \approx 1.54 \times 10^8$ km, and $R_\odot = 381 \times 1756 \approx 6.69 \times 10^5$ km.

Solution to Exercise 5.37

Consider the family of circles which have AB as a chord. Then the centers of these circles lie on the perpendicular bisector of the line segment AB. Let P be the point of tangency of the circle which is tangent to the sloping floor.

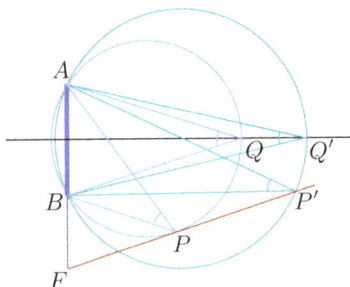

To show that P maximizes the viewing angle, we proceed as follows. Let P' be a point on the right of P as shown, and consider the circle which passes through P', A, B. From the picture above, $\angle APB = \angle AQB > \angle AQ'B = \angle AP'B$.

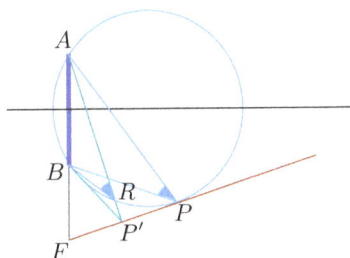

On the other hand, if P' is a point on the left of P as shown above, then we see that $\angle AP'B = \angle ARB - \angle P'BR < \angle ARB = \angle APB$. This completes the proof.

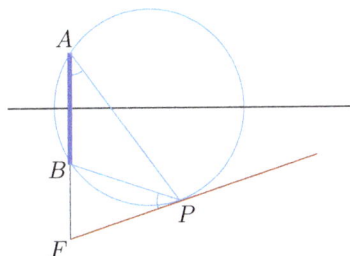

Location of P: Note that $\Delta BFP \sim \Delta PFA$. So $BF/PF = PF/AF$, that is, $PF = \sqrt{AF \cdot BF}$. By knowing the heights from the floor of the top and the bottom of the screen, we can locate the position of P.

Solution to Exercise 5.38

Denote the incenter of ΔABC, right angled at C, by I, and its circumcenter by O. Drop perpendiculars from I to the three sides, meeting them at L, M, N.

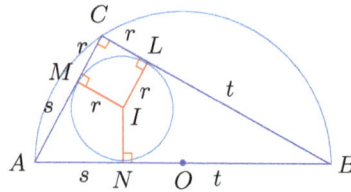

Then $CMIL$ is a rectangle, with equal adjacent sides IM, IL, both equal to r. Hence $CMIL$ is a square, with $CM = CL = r$. Set $s := AM = AN$, and $t := BN = BL$. So $AC + CB = (s+r) + (r+t) = 2r + (t+s) = 2r + AB = 2r + 2R$.

Solution to Exercise 5.39

$\angle PT_1O = 90°$, and so $\angle PT_1T_2 = 90° - \angle OT_1T_2$. But since $PT_1 = PT_2$, ΔT_1PT_2 is an isosceles triangle. Hence

$$\angle T_1PT_2 = 180° - 2\angle PT_1T_2 = 180° - 2(90° - \angle OT_1T_2) = 2\angle OT_1T_2.$$

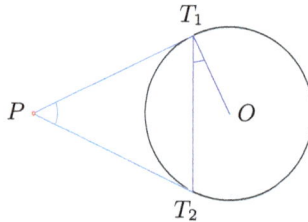

Solution to Exercise 5.40

First consider the case when $AB \parallel \ell$. Construct the perpendicular bisector ℓ' of AB, and suppose that it meets ℓ in T. Draw the perpendicular bisector of BT, and suppose it meets ℓ' at O. Then O is the circumcenter of ΔABT, and this is the required circle, since it passes through A, B and is tangential to ℓ at T (because the angle made by ℓ with the radius OT at T is $90°$).

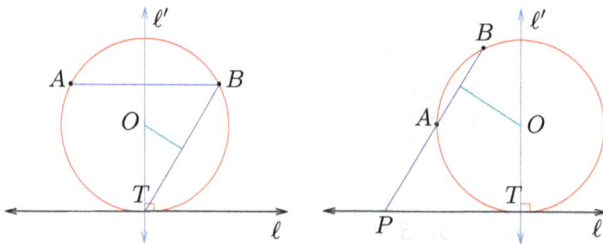

Now let us consider the case when AB is not parallel to ℓ. Denote by P the point of intersection of AB extended with the line ℓ. Knowing the lengths PA, PB, we can construct their geometric mean $\sqrt{PA \cdot PB}$ as follows: Draw a line segment XY of length $PA + PB$. Let Z be a point on XY such that $XZ = PA$. Draw a circle with XY as the diameter, and let W be one of the two points of intersection of the perpendicular to XY at Z and the circle. Then by Exercise 5.7, we know that $ZW = \sqrt{XZ \cdot ZY} = \sqrt{PA \cdot PB}$. Now with P as center and radius ZW, draw a circle, and suppose that it cuts ℓ at a point T. (There will be two points of intersection with ℓ, and we may take any one of them as T.) Erect a line ℓ' passing through T which is perpendicular to ℓ. Construct the perpendicular bisector of AB, and suppose that it meets ℓ' at O. With center O and radius OA, draw a circle. This is the required circle.

Solution to Exercise 5.41

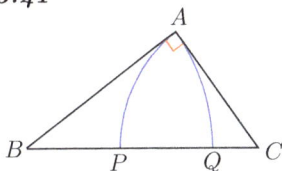

We have $AB^2 = BP(BP + 2AC)$ and $AC^2 = QC(QC + 2AB)$.

Using Pythagoras's Theorem, $BC^2 = AB^2 + AC^2$, and so we obtain that $(BP + PQ + QC)^2 = BC^2 = BP(BP + 2AC) + QC(QC + 2AB)$. This yields

$$PQ^2 = 2(AC - PQ)BP + 2QC(AB - PQ) - 2BP \cdot QC$$
$$= 2QC \cdot BP + 2QC \cdot BP - 2BP \cdot QC = 2QC \cdot BP.$$

Notes

The proof without words in Exercise 3.21 is based on [Kawasaki (2005)].
The intuitive explanation given on page 252 is taken from [Tokieda (2014)].

Bibliography

Alsina, C. and Nelsen, R. (2015). Proof Without Words: President Garfield and the Cauchy-Schwarz Inequality. *Mathematics Magazine* **88**, pp. 144–145.

Bogomolny, A. (1997). *Cut the Knot*, mathematical web resources available at `www.cut-the-knot.org/`

Bryant, J. and Sangwin, C. (2008). *How Round is Your Circle? Where Engineering and Mathematics Meet* (Princeton University Press).

Courant, R. and Robbins, H. (1979). *What is Mathematics? An Elementary Approach to Ideas and Methods* (Oxford University Press).

Dolan, S. (2011). Fermat's Method of "Descente Infinie". *Mathematical Gazette* **95**, pp. 269–271.

Frohliger, J. and Hahn, B. (2005). Honey, Where Should we Sit? *Mathematics Magazine* **78**, 5, pp. 379–385.

Gutenmacher, N. and Vasilyev, A. (1980). *Straight Lines and Curves* (Mir Publishers).

Hess, A. (2012). A Highway from Heron to Brahmagupta. *Forum Geometricorum* **12**, pp. 191–192.

Kawasaki, K. (2005). Proof Without Words: Viviani's Theorem. *Mathematics Magazine* **78**, p. 213.

Kocik, J. (2012). Geometric Diagram for Relativistic Addition of Velocities. *American Journal of Physics* **80**, p. 737.

Moreno, S. and Garcia-Caballero, E. (2013). Entry 1.414 $\cdots$ in Ramanujam's Notebooks: $\sqrt{2}$ is irrational. *Mathematical Gazette* **97**, 539, p. 329.

Sasane, A. (2015). *The How and Why of One Variable Calculus* (Wiley).

Shirali, S. (1998). On Kepler's First Law: The Law of Ellipses. *Resonance* **3**, 5, pp. 30–42.

Tokieda, T. (1998). Mechanical Ideas in Geometry. *American Mathematical Monthly* **105**, 8, pp. 697–703.

Tokieda, T. (2014). *Topology and Geometry*. Lecture videos available on the web at `www.youtube.com/user/AIMSacza/playlists`.

Tong, D. (2012). *What Would Newton Do?* Talk slides, available on the web at `www.damtp.cam.ac.uk/user/tong/talks/newton.pdf`.

Trigg, C. (1985). *Mathematical Quickies* (Dover).

Vaidya, A., Rao, K. and Singh, U. (1988). *Mathematics: A Textbook for Secondary Schools. Class IX* (NCERT).

Yaglom, I. (1979). *A Simple Non-Euclidean Geometry and its Physical Basis. An Elementary Account of Galilean Geometry and the Galilean Principle of Relativity* (Springer).

Index